Low Energy Flight: Orbital Dynamics and Mission Trajectory Design

Jianping Yuan · Yu Cheng ·
Jinglang Feng · Chong Sun

Low Energy Flight: Orbital Dynamics and Mission Trajectory Design

Science Press
Beijing

Springer

Jianping Yuan
School of Astronautics
Northwestern Polytechnical University
Xi'an, China

Yu Cheng
School of Astronautics
Northwestern Polytechnical University
Xi'an, China

Jinglang Feng
Nanjing University
Nanjing, China

Chong Sun
School of Astronautics
Northwestern Polytechnical University
Xi'an, China

ISBN 978-981-13-6129-6 ISBN 978-981-13-6130-2 (eBook)
https://doi.org/10.1007/978-981-13-6130-2

Jointly published with Science Press, Beijing, China
The print edition is not for sale in China Mainland. Customers from China Mainland please order the print book from: Science Press.

Library of Congress Control Number: 2019930644

This Springer imprint is published by the registered company Springer Nature Singapore Pte Ltd.
The registered company address is: 152 Beach Road, #21-01/04 Gateway East, Singapore 189721, Singapore

Preface

In recent years, space missions with the destinations to planets and small solar system bodies have become more and more popular. For missions to such kind of celestial bodies, one challenge comes from the trajectory design and optimization of mission transfer that requires low energy, e.g., utilizing the continuous low thrust and invariant manifold. The other challenge is the strong perturbation on the spacecraft's motion from the highly irregular gravity field. This book provides an overview of the major issues related to the development of low-energy flight and includes the continuous low-thrust transfer between the near earth asteroids using the solar sailing and solar electrical propulsion systems that are covered in Chap. 2, the low-energy transfer between libration point orbits and lunar orbits using invariant manifold in Chap. 3, the Lorentz force formation flying under artificial magnetic field in Chap. 4, and the highly non-linear dynamical environment in the vicinity of small solar system bodies that are covered in the Chaps. 5 and 6. Therefore, this book is suitable for both graduate students and researchers.

The contents of this book are based on the recent research and Ph.D. studies of the authors, i.e., Prof. J. Yuan of Northwestern Polytechnic University (NWPU), Dr. C. Sun from 2013 to 2015 of NWPU that contributes Chap. 2, Dr. Y. Cheng from 2014 to 2016 as a joint Ph.D. student at Universitat de Barcelona and NWPU contributing to Chaps. 3 and 4, and Dr. J. Feng from 2011 to 2016 of Delft University of Technology for Chaps. 5 and 6.

In addition, the authors got valuable comments and suggestions from colleagues and friends, and they express their deep gratitude toward Prof. X. Hou from Nanjing University; Prof. G. Gómez, Prof. J. Masdemont, and Prof. À. Haro from Universitat de Barcelona; and Ir. Ron Noomen, Prof. B. Ambrosius, Prof. B. Vermeersen, and Prof. P. Visser from Delft University of Technology.

Xi'an, China
March 2018

Jianping Yuan

Contents

Chapter 1
Introduction

The Sputniks 1 satellite, launched in 1957, was a milestone mission and marked the start of human's journey to space. Driven by our stronger dreams to explore the deep space and with the aid of the fast-developing technologies, human's activities have extended to the space far away from Earth, with destinations including the Moon, the planets and small bodies in our solar system. Numerous missions have been launched. The one that travels the furthest from Earth is the Voyager 1 spacecraft (NASA), which obtained images of Saturn and Jupiter during their flybys. Currently, it has arrived at the outer solar system, i.e. the edge of the solar influence and interstellar medium.

For the future deep space missions, the first challenge comes from the limited amount of available propellant, since the flight destination is far from Earth and the flight time is resultantly longer than Earth missions. Traditionally, the high-impulsive chemical propulsions are used to change the velocity of the spacecraft. For instance, the Hohmann transfer, based on the two-body problem, has been used for transfer trajectory design. Although these techniques have been applied successfully in many space missions, the most energy-efficient means should be addressed to handle the new rising challenge.

Secondly, the mission requirements and flight environment will become more and more complicated. For spacecraft around Earth, the forces in addition to the Earth's central gravity are viewed as small perturbations, which have been handled with mature techniques based on the well-developed perturbed two-body problem. However, in deep space, the gravitational forces from other celestial bodies are large and even comparable with that of Earth. One example is the spacecraft around the collinear libration points of the circular restricted three-body problem (CRTBP) such as the Sun-Earth system, where both primary bodies exert forces of approximately the same magnitude.

Thirdly, the destinations will be even more diverse, i.e. not restricted to large and spherical planetary bodies. Small size and non-spherical bodies, e.g. small solar system bodies including asteroids and comets, have become popular targets. The

J. Yuan et al., *Low Energy Flight: Orbital Dynamics and Mission Trajectory Design*, https://doi.org/10.1007/978-981-13-6130-2_1

explorations of them have significant scientific values, as they are recognized as the remnants of the early solar system and contain rare materials that are potential energy sources. For the spacecraft orbiting around highly irregular small bodies, perturbations from the large non-spherical terms of these bodies make the dynamics and the design of the spacecraft trajectory completely different from those of the two-body problem. Therefore, efforts should be made to design orbits with these large perturbations.

In these scenarios, the classical theory of astrodynamics based on the two-body problem is no longer valid to describe and solve the orbital dynamics. Therefore, the new kinds of propulsions, more accurate dynamical descriptions, and new techniques that fully utilize the natural forces and inherent dynamics, are required to achieve the low-energy flight, which is the focus in this book. These requirements bring new challenges to mission design, such as efficient analytical/numerical ways to study the orbital dynamics with low thrust, new methods to deal with strong perturbations, new ways of designing optimal trajectories for transfer, station-keeping and formation flying, etc. Therefore, modern astrodynamics needs to be applied, which basically includes but not limited to the three-body problem, the dynamical system theory, the low-thrust trajectory design, and orbital dynamics around highly irregular bodies.

For each space mission, mission orbits design mainly consists of two parts: the first is the design of the trajectory transferring from Earth to the mission orbit around the target celestial body; the second part is designing the mission orbits for scientific observations and experiments. Therefore, the analysis of mission orbit design includes the following two major concerns:

- How does the spacecraft been transferred to the target orbit?
- How can the spacecraft be kept on the target orbit or be transferred to new mission orbits for difference mission scenarios or for mission extensions?

This book aims at investigating the above problems in the concept of low-energy flight and in the framework of modern astrodynamics, using techniques such as optimization, perturbation methods and dynamical tools for instance periodic orbits, invariant manifolds, etc. The following three topics are the focuses:

- Low-thrust transfer trajectory design and formation flying;
- Transfer in the Earth-Moon system using invariant manifolds;
- Orbital dynamics around highly irregular asteroids.

1.1 Low-Thrust Propulsion

There are several ways to generate low thrust, e.g. solar power, electric energy, etc. The electric propulsion is the most popular one, which uses electric or electromagnetic energy to produce continuous low thrust. It has been well developed and widely used in space missions, such as the orbit raising and station-keeping of geostationary satellites. Compared with the chemical propulsion, it is more fuel efficient, which generates higher speed through the weaker thrust but much longer execution time. Russian satellites have used electric propulsions for decades. In 2001, with electric propulsion, NASA's Deep Space 1 spacecraft performed interplanetary flight for more than 1000 h. In 2002, Japan launched its first electric-propellant GEO satellite Kodama. In 2006, SMART-1 was ESA's first solar-electric propulsion spacecraft to the Moon and ended with a Moon impact. The Dawn spacecraft, launched in 2007, was the first dual asteroid mission actuated by ion thruster. By 2013, approximately 350 spacecraft have flown using electric propulsions.

Chapter 2 is devoted to the optimal transfer trajectory design using the low thrust. A new efficient method called virtual gravity field is proposed to provide initial guesses for the optimization process. This method is demonstrated by several interplanetary transfer examples and compared with other optimization methods.

Apart from the transfer trajectory, low thrust can also be used for keeping mission orbits, e.g. formation flying. As an alternative of the single spacecraft mission, formation flying provides the advantages of more flexibility for maintenance, upgrade, and replacement of part of the formation. There are already some missions used this technique, for instance, NASA's Magnetospheric Multiscale Mission (MMS), launched in 2015, investigating the interaction of the magnetic fields of Sun and Earth. ESA's Laser Interferometer Space Antenna (LISA) mission is going to use a triangular formation in the orbit trailing the Earth in the heliocentric frame. Recently, a new-concept propellant-less formation flying utilizing the Lorentz force is proposed. The leader provides the magnetic field, and the motion of the charged follower in this field generates the Lorentz force. Although it is currently in the research stage, this technique demonstrates a promising application on future space explorations. In spite of some preliminary analysis on the dynamics of this system, Chap. 4 of this book provides systematic studies from different aspects, including equilibrium points, periodic orbits, 2-dimensional invariant tori of three specific cases.

1.2 Transfer in the Earth-Moon System

For missions around major celestial bodies such as Earth, the gravity of the central body is the dominant force, and other forces can be treated as small perturbations. The resulting trajectories are perturbed Keplerian motions that are close to the unperturbed ones. In deep space missions, the circular restricted three-body CR3BP is usually applied to model the orbital motion of the spacecraft, instead of the perturbed two-body problem. For this dynamical system, there are five libration points, around which the dynamical structures such as periodic orbits and invariant manifolds are often used in designing low-energy transfer trajectories in these systems. ISEE-3, launched in 1978, was the first mission utilizes the invariant manifolds to transfer to the first Lagrangian point (L1) of the Sun-Earth system. Launched in 2001, the Genesis mission took the heteroclinic connections between orbits around different collinear libration points as the transfer orbit for the first time. In 2009, NASA's ARTEMIS mission send two spacecraft to the Moon through the lunar ballistic transfer trajectories, which utilized the invariant manifolds of both the Sun-Earth and the Earth-Moon systems. During the first mission extension after departing from the lunar orbit, China's Chang'e-2 spacecraft arrived the second Lagrangian point (L2) of the Sun-Earth system for the first time ever in 2011. The following GRAIL mission, launched in 2011, uses similar transfer trajectories. Launched in 2014, China's CE-5/T1 mission also visited the L2 point of Earth-Moon system from Earth via a lunar swing-by and returned to the Moon through the invariant manifolds associated with this L2 point.

Apart from the CRTBP, there are some improved models such as the bi-circular model, the Hill model, the restricted four-body problem. Nevertheless, only the mostly applied CR3BP will be discussed in this book. Focusing on the Earth-Moon system and using the CR3BP, Chap. 3 is devoted to studying the transfer from a halo orbit to a lunar polar orbit. By varying the amplitude of the halo orbit, the specific unstable orbit in the unstable invariant manifold, the specific point on this specific unstable orbit, and the height of the lunar polar orbit, this problem is systematically studied.

1.3 Orbital Dynamics Around Irregular Bodies

As mentioned, in addition to planets, small solar system bodies with irregular gravity field are also interesting targets for both scientific and technical objectives and have become popular destinations for space missions. In 2000, NEAR (NASA) spacecraft arrived at asteroid 433 Eros and determined its gravity, mass, spin rate and orientation, density and internal mass distribution. In 2005, Hayabusa (JAXA) characterized Itokawa's surface and shape thoroughly. Regolith samples were collected and returned to Earth for the first time in 2010. After ten years' journey, Rosetta (ESA) had a rendezvous with its target comet 67P/Churyumov–Gerasimenko in 2014 and released the lander Philae for the first landing on a comet ever. The comet was revealed to have a highly irregular shape of a contact binary body. For missions to such kind of bodies, one of the biggest challenges comes from the perturbation on the spacecraft's motion from the highly irregular gravitational field. Moreover, due to the weak gravity field of the small bodies, the motion of the spacecraft is more sensitive to perturbations, for instance, solar radiation pressure (SRP) and outgassing etc.

Focusing on the non-spherical perturbation of the asteroids, Chap. 5 is devoted to the ground-track resonances, which is the main cause of the instability of orbital motions close to the asteroid's surface. This gives the readers an intuitive picture of "haotic" orbital environment around these highly irregular objects. After that, Chap. 6 focuses on families of periodic orbits and their associated stability properties. This study provides a powerful way to select stable/unstable mission orbits out of the "haotic sea" around these bodies.

It is stressed here that each chapter is self-contained, since they belong to relatively different sub-topics of modern astrodynamics. The reason that we put Chap. 4 on low-thrust after Chap. 3 is the system and the associated dynamical structures are similar with those of the CR3BP studied in Chap. 3. The readers can better understand the contents of this chapter after they read Chap. 3.

Chapter 2
Continuous Low Thrust Trajectory Design and Optimization

2.1 The Virtual Gravity Field Method

2.1.1 Introduction

Continuous low thrust propulsion is an effective way for achieving space mission trajectory design, which has attracted much attention in literatures. The fundamental task of trajectory design is to find thrust profile which can change spacecraft from one state to another state within a given flight of time. However, the trajectory design is still challenging, because there are too many trajectory parameters which need analyzing and those parameters associated search space is very large.

Analytical solutions have been proposed by several authors. Tsien [21], Boltz [6] and Mengali [16] developed analytical solutions of orbit motion, under the assumption of continuous thrust aligning along the radius direction. Following the same formulation, Boltz [7] and Zee [25], studied the case of tangential thrust for continuous low thrust trajectories. Furthermore, Gao [8] presented an averaging technique to obtain analytical solution in case of tangential thrust. Although those methods can, to a large extent, simplify the continuous low thrust problem, they are only suitable for some special cases.

For general cases, there is no analytical solution in continuous thrust trajectory, and the low thrust trajectory is designed typically using two optimization methods: the indirect optimization method and the direct optimization method. The former is based on the calculus of variation, and then the optimization problem is modeled as a two-point boundary value problem. However, it is extremely sensitive to the initial guess, so it is difficult to generate suitable solutions using the indirect method. The latter parameterizes a trajectory using a few variables, and then the nonlinear programming technique is used to optimize those variables to maximize objective function. A variety of methods of this type have been examined [5]. A noticeable direct method was proposed by Jon A. Sims et al. [11]. In his work, the trajectory was divided into several segments at discrete points. The continuous low thrust trajectory design problem was modeled into a nonlinear programming problem, and was solved

J. Yuan et al., *Low Energy Flight: Orbital Dynamics and Mission Trajectory Design*, https://doi.org/10.1007/978-981-13-6130-2_2

by the nonlinear programming software SNOPT. Another method that can provide an initial guess for more accurate optimizers is the shape-based method developed by [2]. In the shape-based method, the thrust is assumed to be aligned along the velocity direction, and the radical vector of spacecraft is written as a function of the transfer angle. Then the coefficients of the function are calculated to satisfy the boundary constraints. Later, [1] extended the shape-based approximation method, in order to reduce the required thrust to satisfy the thrust constraint. In their method, the radius vector and polar angle are described as functions of flight time in form of Fourier series. Those coefficients of Fourier series are optimized using Fmincon tools in Matlab software to satisfy the boundary constraints and thrust constraints.

In this section, a virtual central gravitational field method (VCGF) is proposed to determine continuous thrust trajectory. Instead of providing some special initial guesses, the method can generate a large number of feasible initial guesses efficiently and find the optimal one. There is no prior assumption about the direction of thrust used in this method, and the solutions are analytically determined by the virtual gravity. The basic idea of the method is that, without thrust and neglecting perturbation, the spacecraft flies in a conic orbit in two-body gravitational field. Similarly, if the thrust and the Earth gravity can form a virtual central gravitational field, then the spacecraft can fly in a virtual conic orbit to accomplish trajectory maneuver in that virtual gravity. In this way, feasible continuous thrust trajectories can be parameterized, and expressed as a kind of displaced orbits named virtual conic orbits analytically. Combined with the Particle Swarm Optimization (PSO) algorithm, the proposed method provides a new way to obtain initial guess given an objective function. This method is not only intended for rendezvous case, but it is also used to solve less constrained cases like orbit interception.

2.1.2 The Definition of Virtual Central Gravitational Field

In the two-body problem, spacecraft flies in a Keplerian orbit in the geocentric gravitational field with no thrust. By analogy, if the continuous thrust and the Earth's gravity can form a virtual central gravitational field, the spacecraft can fly in a virtual Keplerian orbit. A virtual central gravitational field can be defined by two parameters: the magnitude of virtual gravity μ_{vg}, the displaced position parameter r_0, as shown in Fig. 2.1. From the definition of the virtual central gravitational field, we can see that the Earth's gravity is a special case of a virtual gravity, whose gravity parameter is $\mu_{vg} = 1\ DU/TU^2$, $\boldsymbol{r}_0 = 0\ DU$ (here DU is the distance Unit, and the TU is the time Unit).

As shown in Fig. 2.2, point A is an initial point and point B is a final point; r_a, r_b are position vectors in point A and point B in geocentric coordinate system; r_{vga},r_{vgb} are position vectors in point A and point B in the virtual central gravitational coordinates system. Virtual Keplerian orbits can be classified into three types as shown in Fig. 2.2.

(1) $\mu \neq \mu_{vg}$, $r_0 = 0$ as shown in Fig. 2.2a;
(2) $\mu \neq \mu_{vg}$, $r_0 \neq 0$ as shown in Fig. 2.2b;

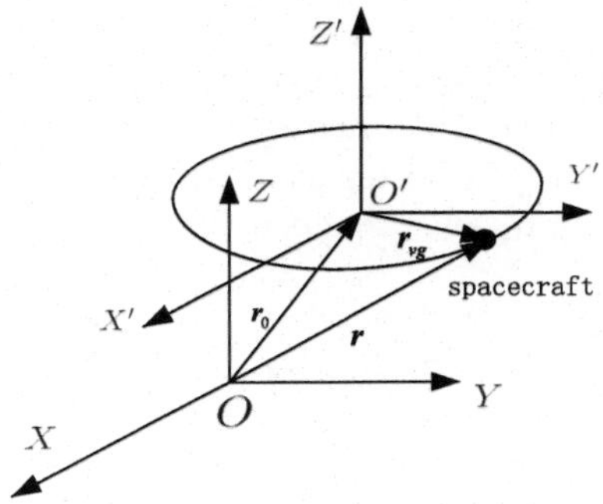

Fig. 2.1 The virtual central gravitational field

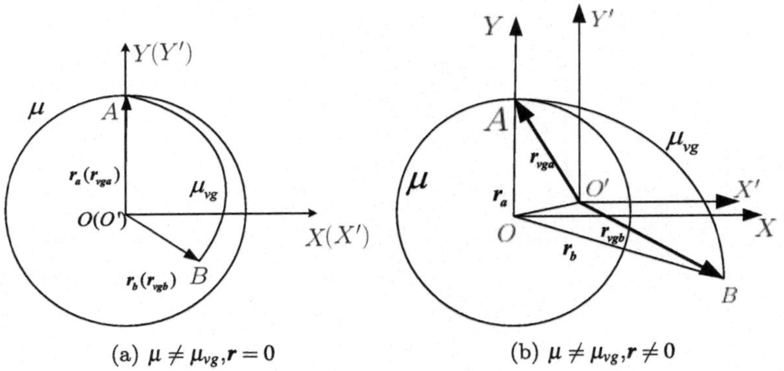

Fig. 2.2 Three types of virtual Keplerian orbits

The VCGF method requires that a spacecraft should fly in a virtual conic orbit and in the virtual central gravitational field to satisfy the boundary constraints. There are a few steps in the VCGF method. Firstly, feasible virtual gravity field determined by parameter set (μ_{vg}, r_0) are required, in which the spacecraft runs in the virtual conic orbits to satisfy trajectory constraints. Secondly, the required thrust to form the virtual gravity is computed, and the fuel consumption is calculated. Finally, considering the objective function, the PSO algorithm is adopted to find the optimal initial guess for more accurate optimizer.

2.2 Trajectory Design Using the VCGF Method

2.2.1 Rendezvous Trajectory Design Using VCGF Method

2.2.1.1 Trajectory Parameterization

In the geocentric coordinate system, r_a and v_a are the position vector and the velocity vector at point A, r_b and v_b are the position vector and the velocity vector at terminal point B. (μ_{vg}, r_0) are parameters of the virtual gravity. In the virtual central gravitational coordinate system, r_{vga} and v_{vga} are the position vector and the velocity vector at point A respectively; r_{vgb} and v_{vgb} are the position vector and the velocity vector at point B respectively. Velocity vectors and position vectors of the spacecraft at initial point A and terminal point B can be computed as,

$$\begin{cases} \boldsymbol{r}_{vga} = \boldsymbol{r}_a + \boldsymbol{r}_0 \\ \boldsymbol{v}_{vga} = \boldsymbol{v}_a \\ \boldsymbol{r}_{vgb} = \boldsymbol{r}_b + \boldsymbol{r}_0 \\ \boldsymbol{v}_{vgb} = \boldsymbol{v}_b \end{cases} \tag{2.1}$$

With the Keplerian orbit theory, one has that,

$$\begin{cases} \boldsymbol{h}_{vg} = \boldsymbol{r}_{vga} \times \boldsymbol{v}_{vga} = \boldsymbol{r}_{vgb} \times \boldsymbol{v}_{vgb} \\ \Delta f = \arccos(\frac{\boldsymbol{r}_{vga} \cdot \boldsymbol{r}_{vgb}}{r_{vga} \cdot r_{vgb}}) \\ r_{vga} = \frac{{h_{vg}}^2}{\mu_{vg}} \frac{1}{1+e_{vg}\cos f_{vga}} \\ f_{vgb} = f_{vga} + \Delta f \\ r_{vgb} = \frac{{h_{vg}}^2}{\mu_{vg}} \frac{1}{1+e_{vg}\cos f_{vgb}} \end{cases} \tag{2.2}$$

where $\boldsymbol{h}_{vg}$, e_{vg} are the angular of momentum and the eccentricity of the virtual Keplerian orbit, while f_{vga}, f_{vgb} are the true anomaly of point A and B in the virtual central gravitational field respectively. The range of parameter f_{vga} is $[0, 2\pi]$, while $\boldsymbol{r}_0$ should satisfied,

$$\boldsymbol{h}_{vg} = \boldsymbol{r}_{vga} \times \boldsymbol{v}_{vga} = \boldsymbol{r}_{vgb} \times \boldsymbol{v}_{vgb} \tag{2.3}$$

In the X-Y plane, we set

$$\boldsymbol{r}_a = [r_{ax}, r_{ay}], \boldsymbol{r}_b = [r_{bx}, r_{by}]$$

,

$$\boldsymbol{v}_a = [v_{ax}, v_{ay}], \boldsymbol{v}_b = [v_{bx}, v_{by}]$$

and then Eq. 2.3 can be expressed as follow,

$$a_1 \cdot r_{0x} + b_1 r_{0y} = c_1 \tag{2.4}$$

Where a_1, b_1, c_1 are expressed as,

$$\begin{cases} a_1 = v_{ay} - v_{by} \\ b_1 = v_{bx} - v_{ax} \\ c_1 = (r_{bx}v_{by} - r_{by}v_{bx}) - (r_{ax}v_{ay} - r_{ay}v_{ax}) \end{cases} \tag{2.5}$$

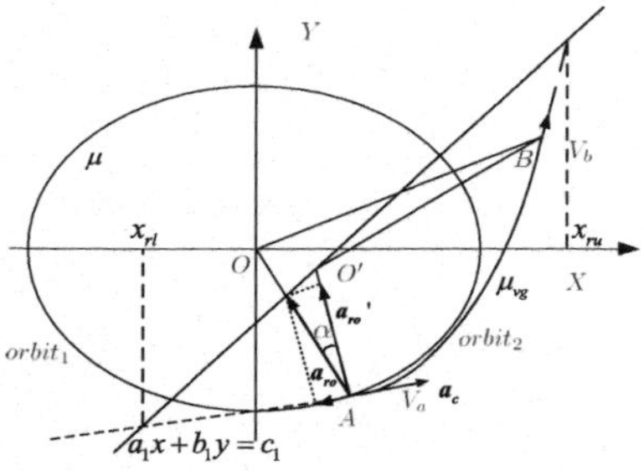

Fig. 2.3 Orbit rendezvous in 2-D space

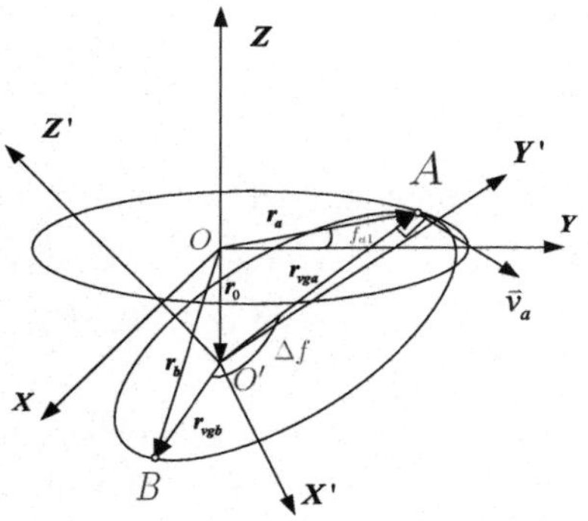

Fig. 2.4 Orbit rendezvous in 3-D space

In the 2-dimensional plane, through Eq. 2.4, it can be obtained that vector $\boldsymbol{r}_0 = [r_{0x}, r_{0y}]$ is located on a line, because the intersection angle between v_{vga} and v_{vgb} is smaller than π, the range of r_0 can be determined as $r_{0x} \in (x_{rl}, x_{ru})$, as shown in Fig. 2.3.

Once parameters (r_{0x}, f_{vga}) are given, $(\boldsymbol{r}_0, \mu_{vg})$ can be calculated using Eqs. 2.2 and 2.4. Each parameter set $(\boldsymbol{r}_0, \mu_{vg})$ can determine a virtual gravity. Therefore, all those feasible trajectories can be parameterized as virtual conic orbits (Fig. 2.3), and analytically expressed as virtual conic orbits. The process of finding rendezvous trajectory for 3-D space is similar as that for 2-D space, as shown in Fig. 2.4.

In the VCGF method, the transfer angle for one single virtual conic orbit is less than 2π, because the spacecraft returns to the initial state after one revolution in the virtual gravity. For multi-revolutions trajectory design problem, the orbit patching technique discussed in [20] is used. The whole trajectory is divided into a few segments at discrete points, and those discrete points are target points or just the control points. Each segment is a virtual conic orbit, and the continuous low thrust trajectory design can be transformed into multi-segments patching problem. More detail can be obtained in Ref. [20].

2.2.1.2 Trajectory Optimization

Through the Keplerian orbit theory, the flight time of a spacecraft can be expressed as follows,

$$f_t = f(f_{vga}, \boldsymbol{r}_{0x}) = \int_{f_{vga}}^{f_{vgb}} \frac{\sqrt{{a_{vg}}^3(1-{e_{vg}}^2)^3}}{\mu_{vg}} \frac{1}{(1+e_{vg}\cos f)^2} df \tag{2.6}$$

Similarly, the energy consumption can be expressed as,

$$f_e = f(f_{vga}, \boldsymbol{r}_{0x}) = \int_{f_{vga}}^{f_{vgb}} (\boldsymbol{F}_2 - \boldsymbol{F}_1) df \tag{2.7}$$

here set $\boldsymbol{r}$ is the position vector from the spacecraft to the Earth,

$$\boldsymbol{F}_1 = \frac{\mu}{r^2}, \boldsymbol{F}_2 = \frac{\mu_{vg}}{{r_{vg}}^2}, r_{vg} = \frac{{h_{vg}}^2}{\mu_{vg}} \frac{1}{1+e_{vg}\cos f_{vg}} \tag{2.8}$$

In minimum fuel consumption case, the objective function is,

$$f_{\Delta V} = f_{\min}(f_{vga}, \boldsymbol{r}_{0x}) = \int_{f_{vga}}^{f_{vgb}} \left(\frac{\mu_{vg}}{{r_{vg}}^2} - \frac{\mu}{r^2}\right) df \tag{2.9}$$

In this optimization problem, $\boldsymbol{r}_{0x}$ and μ_{vg} are independent variables. Feasible variables $\boldsymbol{r}_{0x}$, μ_{vg} need to be optimized, so as to minimize the flight time or the fuel consumption.

2.2.1.3 Thrust Acceleration

In the geocentric coordinate system, equation of motion of the spacecraft under continuous thrust can be written as,

$$\frac{d^2\boldsymbol{r}}{dt^2} + \frac{\mu}{r^3}\boldsymbol{r} = \boldsymbol{T}_{ac} \tag{2.10}$$

Here we assume that the thrust can form a virtual gravity, and then in the virtual central gravitational coordinate system, the equation of spacecraft motion under continuous thrust can be written as,

$$\frac{d^2\boldsymbol{r}_{vg}}{dt^2} + \frac{\mu_{vg}}{{r_{vg}}^3}\boldsymbol{r}_{vg} = 0 \tag{2.11}$$

where

$$\boldsymbol{r}_{vg} = \boldsymbol{r} - \boldsymbol{r}_0. \tag{2.12}$$

Because $\boldsymbol{r}_0$ is a constant in one virtual gravity, we have $(\boldsymbol{r}_0)' = 0$. The position vector and the velocity vector in the virtual centric coordinate system can be computed by Eq. 2.13,

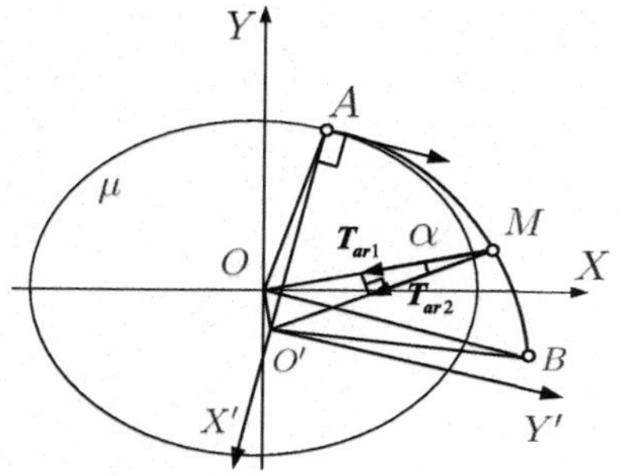

Fig. 2.5 Force analysis in the 2-dimensional space

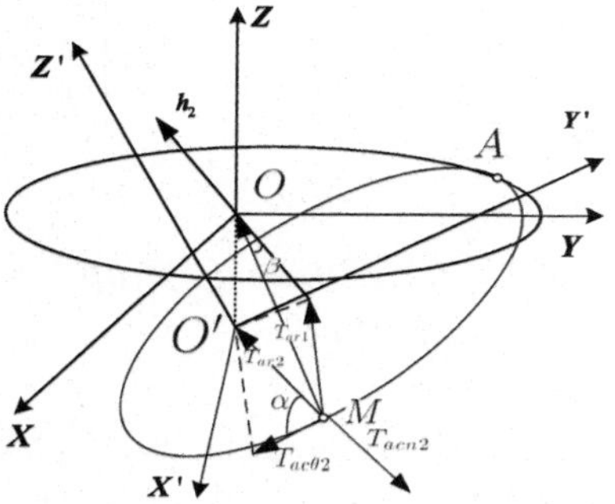

Fig. 2.6 Force analysis in the 3-dimensional space

$$\begin{cases} \boldsymbol{r}_{vg} = \boldsymbol{r} - \boldsymbol{r}_0 \\ \boldsymbol{v}_{vg} = \boldsymbol{v} \end{cases} \tag{2.13}$$

where $\boldsymbol{v}$, $\boldsymbol{v}_{vg}$ are velocity vectors in geocentric coordinates system and virtual central gravitational coordinate system. Through Eqs. 2.10, 2.11and 2.12, the required thrust acceleration (TA) is,

$$\boldsymbol{T}_{ac} = \frac{d^2\boldsymbol{r}}{dt^2} + \frac{\mu}{|\boldsymbol{r}^3|}\boldsymbol{r} - \frac{d^2\boldsymbol{r}_{vg}}{dt^2} - \frac{\mu_{vg}}{|\boldsymbol{r}_{vg}{}^3|}\boldsymbol{r}_{vg} \tag{2.14}$$

As shown in Fig. 2.5, in 2-dimensional space, given $\boldsymbol{r}_0x$, μ_{vg}, the thrust acceleration required in 2-D space can be obtained by Eq. 2.15,

$$\begin{cases} \boldsymbol{r}_{vg} = \boldsymbol{r} - \boldsymbol{r}_0 \\ T_{ar1} = \frac{\mu}{|\boldsymbol{r}|^2} \\ T_{ar2} = \frac{\mu_{vg}}{|\boldsymbol{r}_{vg}|^2} \\ T_{ac} = T_{ar} \cdot \sin\left(\arccos\left(\frac{T_{ar2}}{T_{ar1}}\right)\right) \end{cases} \tag{2.15}$$

where T_{ar1}, T_{ar2}, T_{ac} are magnitude of gravity acceleration, virtual central gravitational acceleration and required TA respectively.

In case of 3-dimensional trajectory design, as shown in Fig. 2.6, $\boldsymbol{r}_M$ is the position vector of spacecraft at point M, and $\boldsymbol{r}_{vgM}$ is the position vector at point M in the virtual gravity. Assuming TA is $\boldsymbol{T}_{ac} = [T_{acn2}, T_{ac\theta 2}]$, where T_{acn2} is aligned along the radical direction, and $T_{ac\theta 2}$ is aligned along the circulation direction; T_{ar1} is the geocentric gravitation force acceleration, and T_{ar2} is the virtual centric gravitational force acceleration; α is the angle between $T_{ac\theta 2}$ and T_{ar2}, and β is the angle between $\boldsymbol{r}_{M1}$ and $\boldsymbol{h}_2$, then the required TA can be computed as,

$$\begin{cases} \beta = \arccos \frac{|\boldsymbol{r}_M \cdot \boldsymbol{h}_{vg}|}{|\boldsymbol{r}_M| \cdot |\boldsymbol{h}_{vg}|} \\ T_{ar1} = \frac{\mu}{r_M{}^2} \\ T_{ar2} = \frac{\mu_{vg}}{r_{vgM}{}^2} \\ \alpha = \arcsin \left(\frac{T_{ar2}}{T_{ar1} \sin \beta} \right) \\ T_{ac\theta 2} = T_{ar2} \sin \beta \cos \alpha \\ T_{acn2} = T_{ar1} \cos \beta \end{cases} \tag{2.16}$$

2.2.2 *Orbit Interception Trajectory Design Using VCGF Method*

The VCGF method can also be applied to designing less constrained interception trajectory. In Eq. 2.2, the unknown variables $\boldsymbol{r}_0, f_{vga}$ or $\boldsymbol{r}_0, \mu_{vg}$ and the virtual conic orbit in Fig. 2.2a are used. In two dimension problem, we can set $\boldsymbol{r}_0 = 0$ to simplify the problem, as shown in Fig. 2.7. Given parameters $\boldsymbol{r}_a, \boldsymbol{r}_b, \boldsymbol{v}_a$, it needs to obtain feasible parameter μ_{vg} to solve Eq. 2.2. While in the three dimension problem, both $\boldsymbol{r}_0, \mu_{vg}$ are required to optimize. A root finding function f-zeros in MATLAB can be used to solve this problem.

In the PSO algorithm, the free parameters are the magnitude of virtual gravity μ_{vg} and the projection of $\boldsymbol{r}_0$ in x axis: $\boldsymbol{r}_0$. The objective function can be the fuel cost of whole trajectory or the flight time. Here we set the maximal number of iteration in PSO algorithm are N. Take the orbit rendezvous trajectory design as an example. Given the range of free variables $r_{0x} \in (r_{0xl}, r_{0xu})$, the initial point $(\boldsymbol{r}_A, \boldsymbol{v}_A)$ and the terminal point B $(\boldsymbol{r}_B, \boldsymbol{v}_B)$.

There is a large amount of parameter set $(\boldsymbol{r}_0, \mu_{vg})$, and each one corresponds to a trajectory, and the feasible trajectories are those can satisfy boundary constraints.

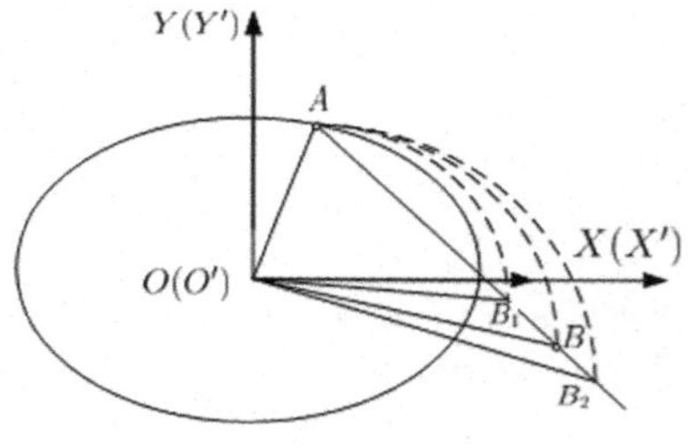

Fig. 2.7 Optimal trajectory using PSO algorithm

The PSO algorithm is adopted to obtain feasible parameter set ($\boldsymbol{r}_0$, μ_{vg}) from their ranges to determine the virtual gravity and the corresponding virtual conic orbits. There are a few steps in the algorithm. To begin with, the range of free parameters is set, and a large number of possible trajectories, same as the number of parameter sets ($\boldsymbol{r}_0$, μ_{vg}), are obtained. Furthermore, those feasible trajectories which can satisfy the constraints are chosen. Finally, the value of the objective function is calculated, and the optimal solution is obtained. The process of PSO algorithm is listed as follow:

Step1: Find the ranges of free variables $\mu_{vg} \in (\mu_{vgl}, \mu_{vgu})$, $r_{0x} \in (r_{0xl}, r_{oxu})$;

Step2: Initialize the particle position with a uniformly distributed random vector (r_{0x}, μ_{vg}) in their ranges;

Step3: Calculate r_{0y} through Eq. 2.4; then the parameter set ($\boldsymbol{r}_0$, μ_{vg}) is obtained;

Step4: Calculate the radius and velocity state of initial point A ($\boldsymbol{r}_{vga}$, $\boldsymbol{v}_{vga}$), and terminal point B ($\boldsymbol{r}_{vgb}$, $\boldsymbol{v}_{vgb}$)in the virtual gravity through Eq. 2.1;

Step5: Calculate transfer angle from point A and point B in virtual gravity field through Eq. 2.2;

Step6: Calculate the orbital elements of point A through its radius and velocity in virtual gravity field, coe_{vgA}=[a_{vgA}, e_{vgA}, i_{vgA}, w_{vgA}, Ω_{vgA}, TA_{vgA}], here TA_{vgA} are the virtual conic orbital inclination, argument of the periapsis, longitude of the ascending node, and true anomaly respectively.

Step7: Calculate the true anomaly at point B in the designed virtual gravity (TA_{vgb}=$TA_{vga} + \Delta\theta$);

Step8: Calculate the radius and velocity $\boldsymbol{r}_{vgb}$, $\boldsymbol{v}_{vgb}$ through the orbital elements, coe_{vgB}=[a_{vgA}, e_{vgA}, i_{vgA}, w_{vgA}, Ω_{vgA}, TA_{vgB}];

Step9: Calculate the radius error $\Delta\varepsilon_r$=$\left|\boldsymbol{r}_{vgB} - \boldsymbol{r}_{vgb}\right|$ and the velocity error $\Delta\varepsilon_v$=$\left|\boldsymbol{v}_{vgB} - \boldsymbol{v}_{vgb}\right|$; if $\Delta\varepsilon_r \leq 10^{-4}$, $\Delta\varepsilon_V \leq 10^{-4}$; the trajectory in the virtual gravity can satisfy the boundary constraint. Then record the parameter set $(\boldsymbol{r}_0, f_{vga})_i$, and calculate the fuel consumption (or the flight time) as its fitness; else if $\Delta\varepsilon_r > 10^{-4}$, $\Delta\varepsilon_V > 10^{-4}$; then set the fitness of this particle as 1;

Step10: Initialize the particle's best-known position to its initial position; then update the swarms best known position and its fitness; If the number of iteration $i < N$; return to step4. If the number of iteration $i > N$, finished; It should be noted that, there is a large number of feasible virtual Keplerian orbits that can satisfy boundary constraints, and the optimal one is the subcategory of them.

2.2.3 *Mission Applications*

In order to verify the effectiveness of the proposed method, three application examples of the VCGF method are presented in this section. The first one is the Earth-Mars orbital transfer trajectory design. The proposed method was used to generate initial guesses of continuous thrust rendezvous trajectory. Those solutions were compared with the shape-based (SB) method given in [24], in terms of the magnitude and the direction of thrust, transfer angle and fuel consumption. The second example is a fuel-optimal Earth-Mars-Ceres flight trajectory mission discussed in the same reference. The whole trajectory consists of an interception trajectory from the Earth to the Mars, and a rendezvous trajectory from the Mars to the Ceres. The effectiveness of the VCGF method is evaluated by providing an initial guess for the direct optimizer. Solutions of the VCGF method are compared with the three-dimension, shape-based method solution discussed in [24]. Here the direct optimizer GPOPS, a Matlab software for solving a nonlinear optimal control problems, is selected to generate an accurate solution based on the proposed method and the shape-based method. The last case is a collision-speed-maximal interception trajectory design problem in Ref. [12]. In this example, a spacecraft is transferred from the Earth to intercept with a hazardous asteroid. The asteroid 99942 Apophis is the potentially hazardous asteroid on an impact trajectory toward the Earth. The aim of optimization is to maximize the relative speed between the spacecraft and the asteroid. For well-documented reasons, heliocentric canonical units were used. In this paper, the distance units (AU) and the time units (TU) are: 1 DU = 1 (AU) = 149596000 km, 1 TU = 1/2 π year = 58.17 days. All of the examples have been performed on an Intel Core 2.6GHz with Windows 8. The computation time of optimization is calculated by MATLAB tic-toc command.

Example A: The Earth-Mars orbit rendezvous

In this example, the two-dimension Earth-Mars orbit is performed. A spacecraft is transferred from the Earth to rendezvous with the Mars. The boundary conditions are listed in Table 2.1. Here, in order to analyze the relationship between the transfer angle, the required thrust and the fuel cost, the VCGF method is applied to generate initial guesses in case of three transfer angles ($\Delta\ \theta$): a half revolution ($\Delta\ \theta = \pi$), one revolution ($\Delta\ \theta = 2\pi$)and two revolutions($\Delta\ \theta = 4\pi$). Using the VCGF method, those feasible trajectories are expressed as virtual conic orbits and parameterized by the parameter set ($\boldsymbol{r}_0$, μ_2). For one virtual conic trajectory, the transfer angle is less than 2π ($\Delta\theta < 2\pi$). While in case of multi-revolutions condition ($\Delta\theta > 2\pi$), the whole trajectory is divided into a few segments at discrete points, and each segment corresponds to a virtual conic orbit. In order to testify the suitability of proposed method, the results are compared with the five-degree inverse polynomial shaped based method, in terms of magnitude and direction of required thrust, the flight time and the fuel consumption. The boundary conditions of the Earth-Mars rendezvous mission are listed in Table 2.1.

The transfer angles of those cases are assumed to be π, 2π, 4π, respectively. The corresponding flight times are set as 3.75TU, 7.683TU and 16.69 TU respectively. Trajectories generated by the VCGF method and the shape-based method are shown in Fig. 2.8 (for the case of n = 2). The required thrust for all three cases is shown in Figs. 2.9, 2.10 and 2.11. Table 2.2 shows the parameters of the virtual gravity

Table 2.1 Initial and final conditions for transfer trajectory

$r_i = 1AU$	$\theta_i = 0$	$\dot{r}_i = 0$	$\dot{\theta}_i = 0.6564$
$r_f = 1.52AU$	$\theta_f = n\pi$	$\dot{r}_f = 0$	$\dot{\theta}_f = 0.5333$

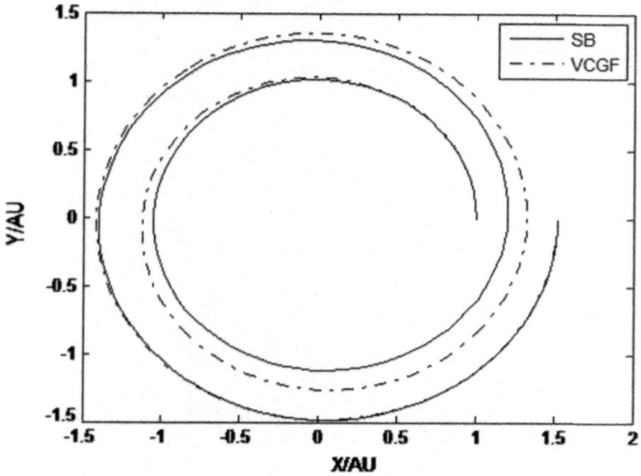

Fig. 2.8 Rendezvous orbits (n = 4)

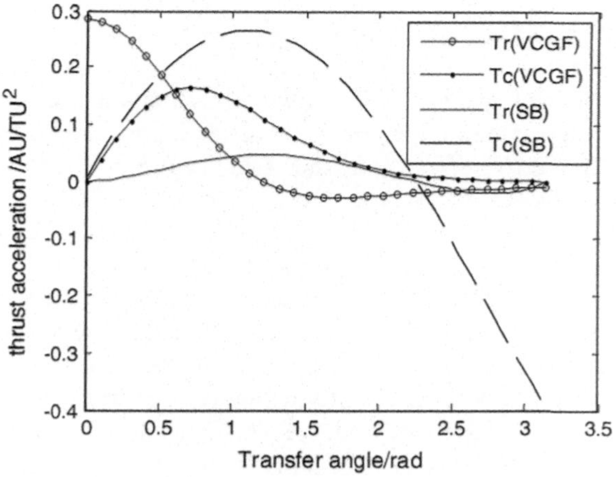

Fig. 2.9 TA for rendezvous orbits (n = 1)

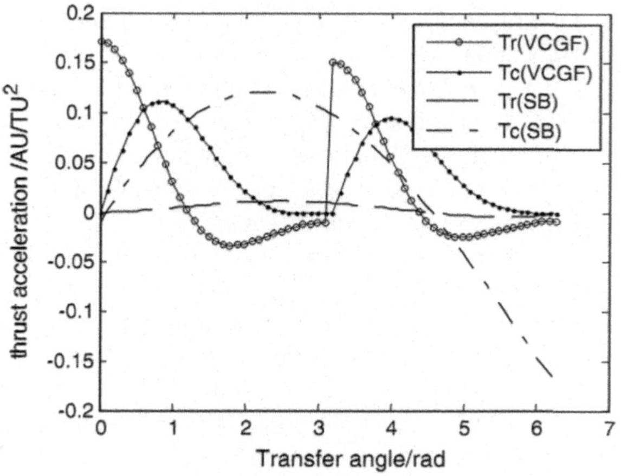

Fig. 2.10 TA for rendezvous orbits (n = 2)

Fig. 2.11 TA for rendezvous orbits (n = 3)

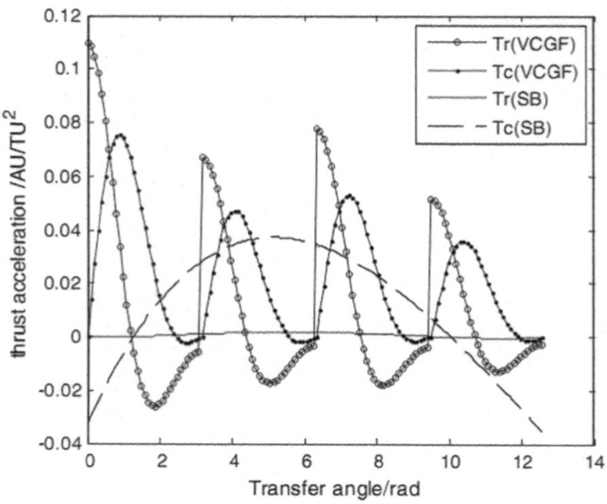

obtained by the VCGF method. The simulation results show that, in the shape-based method, the maximum thrust accelerations are $0.3854AU/TU^2$, $0.1473AU/TU^2$, $0.0372AU/TU^2$; and the fuel costs are $0.638AU/TU$, $0.559AU/TU$ and $0.3688AU/TU$ respectively; For the VCGF method, the maximum thrust accelerations are $0.3561AU/TU^2$, $0.1852AU/TU^2$, $0.1096AU/TU^2$; and corresponding fuel consumptions are 0.6723 AU/TU, 0.6371 AU/TU, 0.6176 AU/TU respectively. From the simulation results we know that there are two main differences between initial guesses generated by the VCGF method and the shape-based method. Firstly, in the shape-based method, the magnitude of thrust and the fuel consumption decrease greatly as increase of transfer angle. Secondly, compared with the shape-based method, the VCGF needs larger thrust and more fuel consumption, especially in the case of long flight time. The main reasons for the differences are that, in the shape-based method, the tangential thrust is consist with large circumferential thrust and small radical thrust in the shape based method, as shown in Figs. 2.9, 2.10 and 2.11. The magnitude of thrust is determined by the transfer angle of trajectory. The mathematical demonstration can be obtained from the dynamics model of shape-based method in [24]. However, in the VCGF method, the key factor affecting the virtual gravity field is the magnitude of virtual gravity and the position in the near circular orbital rendezvous case, but those two parameters are not affected greatly by the increase of the transfer angle. Thus, the required thrust acceleration does not change greatly. Furthermore, the required thrust is consist with higher radical thrust and lower circumferential thrust in the VCGF method. Conversely, the thrust in the shape based method is consist with higher circumferential thrust and lower radical thrust. While compared with orbital maneuver using radical thrust, it is more efficient using circumferential thrust than that of radical thrust in terms of fuel cost, but usually needs more flight time [6]. Thus, a conclusion that can be safely obtained is, the shape-based method is more efficient than the VCGF method in terms of fuel

Table 2.2 Parameters of the VCGF

Transfer angles	r_0	μ_{vg}
π, $(n = 0.5)$	$r_0 = (-0.1286, 0)$	$\mu_{vg} = 0.8188$
$\pi, n = 1$	$r_0 = (-0.05490)$	$\mu_{vg} = 0.9063$
	$r_0 = (-0.0748, 0)$	$\mu_{vg} = 0.8974$
$4\pi n = 2$	$r_0 = (-0.03, 0)$	$\mu_{vg} = 0.9444$
	$r_0 = (-0.025, 0)$	$\mu_{vg} = 0.9573$
	$r_0 = (-0.0387, 0)$	$\mu_{vg} = 0.9415$
	$r_0 = (-0.0362, 0)$	$\mu_{vg} = 0.9504$

Table 2.3 Orbit elements of Mars and Ceres (MJD2000:4748.5)

Planet	a/AU	e	i/deg	Ω/deg	ω/deg	f/deg
Mars	1.524	0.0934	1.8506	348.7422	114.2112	41.990
Ceres	2.7658	0.078	10.607	80.329	72.522	57.3248

cost, especially in case of long flight time rendezvous. Indeed, from the perspective of trajectory shape, the VCGF method can be regarded as a special case of the shape based method, because each arc generated by the VCGF method is a conic, and its shape is determined by variables. But the difference lies in, the required thrust in the VCGF method is not necessary to be aligned along the direction of the spacecraft velocity, and it needs large radical thrust and small circumferential thrust to form the virtual gravity.

Example B: The Earth-Mars Flyby-Ceres flight trajectory

In this example, the proposed method is applied in a more complicated problem given in [12]: the Earth-Mars-Ceres rendezvous mission. The flight trajectory includes an interception trajectory form the Earth to the Mars, and a rendezvous trajectory from the Mars to the Ceres. The spacecraft is given the same launch window: year 1990–2049 and the same launch velocity ranges. The total time of flight constraints: less than 1133 days. The classical orbital elements of the Mars and the Ceres are listed in Table 2.3. The planetary orbital elements corresponding to the date on January 01, 2013 (MJD 2000: 4748.5) are interpreted from the HORIZONS Web Interface of JPL, and the free parameters boundaries are listed in Table 2.4.

The aim of preliminary design using proposed method is to provide a reasonable initial guess to approximately determine the optimal launch time, the time of flight and revolutions to more accurate optimizer. In order to verify the suitability of the VCGF method, the three-dimensional shape-based method proposed in [24] is applied in this mission. After obtaining the initial guess, a direct method GPOPS, is applied to find the optimal result. A total number of 30 nodes are used for the Earth-Mars-Ceres trajectory. The thrust acceleration is assumed constant, $0.2694 DU/TU^2$. In the optimization process, there are seven state parameters $(r, \theta, z, v_r, v_\theta, m)$, and three controls parameters (T_n, T_c, T_r).

Table 2.4 Orbit elements of Mars and Ceres (MJD2000:4748.5)

Optimization parameter	Bounds
Launch date	01/01/1990-12/31/2049
Magnitude of V_∞ at Earth departure	[0.75,200] km/s^2
In plane angle of V_∞ at Earth departure	$[-\pi,\pi]$ rad
Out plane angle of V_∞ at Earth departure	$[-\pi,\pi]$ rad
Time of flight from Earth to Mars	[0,3] years
Magnitude of V_∞ at Mars arrival	[0,3] km/s^2
In-plane angle of V_∞ at Mars arrival	$[-\pi,\pi]$ rad
Out-plane angle of V_∞ at Mars arrival	$[-\pi,\pi]$ rad
Out-of-plane angle of Mars flyby plane	$[-\pi,\pi]$ rad
Time of flight from Mars to Ceres	[0,3] years

Table 2.5 The comparison of preliminary design and accurate optimization

Flight parameters	Shape-based	Optimal result	VCGF	Optimal result
Launch date (yy/mm/dd)	2016/05/29	2016/07/26	2016/07/23	2016/09/17
Arrive date (yy/mm/dd)	2019/07/06	2019/06/30	2019/02/04	2019/09/25
Time-of-flight (day)	1133	1130	927	1103
Final mass/initial mass	0.290	0.252	0.4270	0.3210
CPU time (s)	686.9	1560	653.3	1023.3

The optimal variables in the preliminary stage and optimization stage are shown in Table 2.5. The final optimal Earth-Mars-Ceres trajectory using the VCGF initial guess is shown in Fig. 2.12, and the direction of optimal thrust is shown in Fig. 2.13. From the simulation results, two conclusions can be obtained. Firstly, the optimal result based on the VCGF method is larger than that of the shape-based method in terms of fuel consumption. It is important to point out that, in order to form the virtual gravity field, the required thrust is consist with large radical thrust and small circumferential thrust; hence in terms of fuel cost, it is less efficient than the shape-based method which uses tangential thrust. This result is consistent with the conclusion obtained in Example A. However, it should be noticed that the VCGF method is more efficient than the shape based method in terms of computation time, just as shown in Table 2.5. The main reason lies in the fact that the thrust direction is assumed to be along the tangent direction in the shape based preliminary design. However, the optimal results using the shape-based method are inconsistent with the assumption. Conversely, in the VCGF method, thrust directions are determined by the parameters of virtual gravity field, which is similar as that of optimal result.

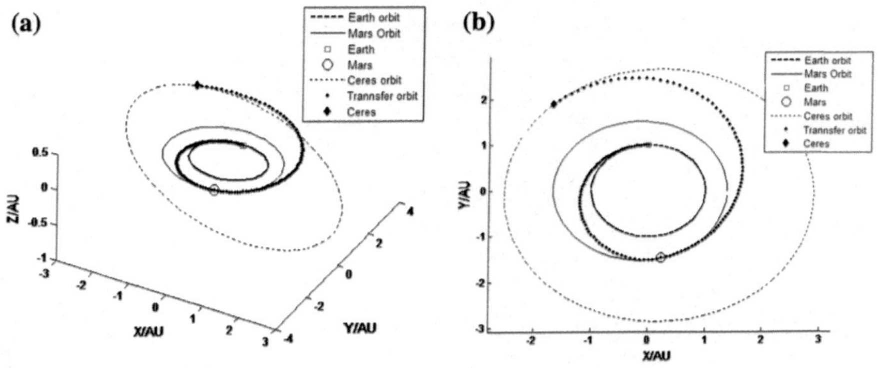

Fig. 2.12 The Earth-Mars-Ceres trajectory

Fig. 2.13 The thrust direction in the Earth-Mars-Ceres trajectory mission

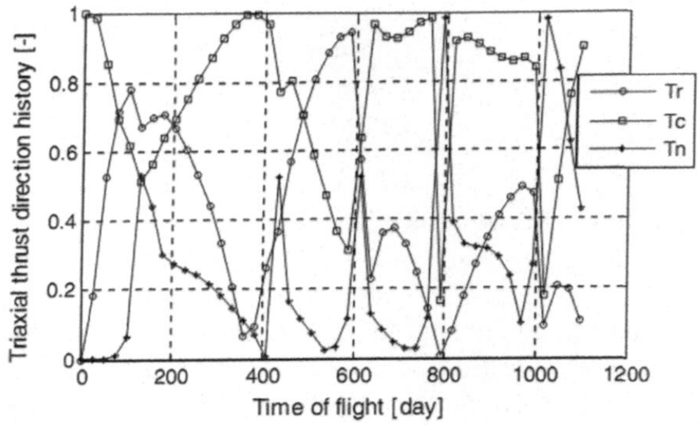

Secondly, the VCGF method can parameterize feasible trajectories using only two variables, so it is simpler to get the initial guess for continuous thrust trajectory.

Example C: Collision-speed, maximal interception trajectory

In order to verify the efficiency of the proposed method, it is applied in the near earth asteroid collision-speed maximal interception trajectory design and optimization problem given in [9]. The relative velocity between the spacecraft and the hazardous asteroid is considered as the performance index, and the aim of optimizing is to maximize it. Here Asteroid 99942 Apophis is assumed to be the target of the potentially hazardous asteroid on an impact trajectory toward the Earth. The launch window is supposed to be located between January 1, 2016 and January 1, 2018. The departure energy is assumed to be no more than 1. The time-of-light is limited to be no more than 800 days, and the search step is set as 5 days. Here the solar-electric propulsion model is utilized, and the performance parameters of solar-electric propulsion engine are given in [9]. The classical orbital elements of the Earth and the Apophis are listed in Table 2.6. The planetary orbital elements corresponding to the date on January 01, 2013 (MJD2000: 4748.5) are interpreted from the HORIZONS Web Interface of JPL.

Table 2.6 Orbit elements of Apophis and Earth (MJD2000:4748.5)

Planet	a/AU	e	i/deg	Ω/deg	ω/deg	f/deg
Earth	0.921	0.191	3.329	204.3498	126.4037	133.7569
Apophis	0.999	0.06063	0.0016	174.2831	287.7551	358.5111

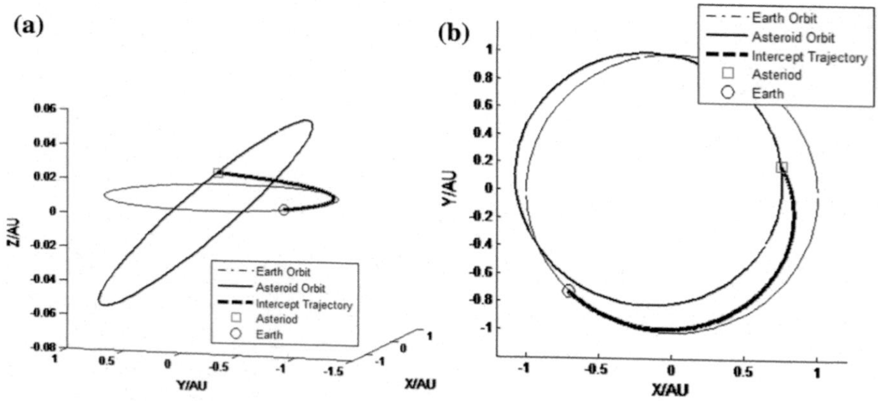

Fig. 2.14 Interception trajectory from the Earth to Asteroid

In this case, the normalized performance index $\frac{v_{frel}}{v_E}$ is considered as fitness. Here, v_E denotes the Earth's revolution speed around the Sun, and v_{frel} denotes the relative speed between the spacecraft and the asteroid. Using the VCGF method, those feasible interception trajectories are expressed as virtual Keplerian orbits in the virtual gravity. A large number of feasible interception trajectories are parameterized, and can be provided as initial guesses. The exponential sinusoid shape based method mentioned in [23] is applied in this mission, to compare the solutions obtained by the VCGF method. Based on the initial guess provided by two analytical approaches, the direct optimizer discussed in [9] is adopted to get the optimal solution. A total number of 30 Legendre-Gauss-Radau (LGR) points are applied in this simulation in order to accurately optimize the Earth-Apophis low thrust trajectory. Then those discrete points generate a nonlinear programming problem, which is solved by the sequential quadratic programming (SQP) method [12]. The direction of optimal thrust based VCGF initial guess is shown in Fig. 2.15, and the final accurate optimization results are shown in Fig. 2.14. The parameters of trajectory for the preliminary stage and the accurate ones are shown in Table 2.7.

The simulation results show that the computation time using the VCGF method is less than that using the shape-based method. The main reason lies in the fact that the engine thrust direction is assumed to be along the tangent direction in the exponential sinusoid shape based method. However, there is no constraint in the VCGF method. Moreover, it can be observed that the fuel consumption of the VCGF method is more than that of the shape based method, but the optimal performance index based on

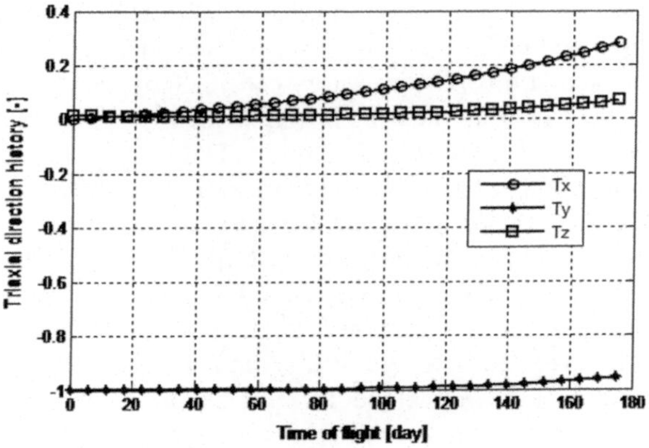

Fig. 2.15 Interception trajectory from the Earth to Asteroid

Table 2.7 The comparison of preliminary design and accurate optimization

Flight parameters	Shape-based	Optimal result	VCGF	Optimal result
Launch date (yy/mm/dd)	2016/05/29	2016/07/26	2016/07/23	2016/09/17
Arrive date (yy/mm/dd)	2019/07/06	2019/06/30	2019/02/04	2019/09/25
Time-of-flight (day)	1133	1130	927	1103
Performance index	0.3572 v_E	0.573 v_E	0.4133 v_E	0.5993 v_E
Final mass/initial mass	0.9385	0.500	0.953	0.5602
CPU time (s)	473.1	1220.9	391.3	1072.1

the VCGF is better than that of the shape based method. The key factor is that, in the VCGF method, the gravity constant is optimized to maximize the relative speed between the spacecraft and the asteroid, so it is flexible to change the speed and position in the radical direction, but it requires large radical thrust to form this kind of virtual gravity. Thus, the perform index using the VCGF method outperforms the shape-based method at the expense of more fuel consumption.

2.2.4 Conclusion Remarks

A novel approach for continuous low thrust trajectory design and optimization problem is proposed. Three types of virtual gravity defined by the magnitude and the position of virtual gravity are presented. Furthermore, the particle swarm optimization algorithm is adopted to obtain a large number of feasible trajectories which can satisfy the constraints. All of those trajectories can be parameterized using parameters of the virtual gravity and analytically expressed as virtual conic orbits. In this way, the non-Keplerian continuous low thrust trajectory can be solved analytically. Instead of providing some special initial guesses for the more accurate optimizers, this method can provide a large number of feasible initial guesses, and generate an optimal one for a variety of spacecraft trajectory design problems. Although energy consumption of the trajectory generated by the proposed method is greater than that of the true solution using the direct method, its computational time is much less than that of the true optimal solution. Therefore, it is an efficient way to provide optimal initial guesses for orbit rendezvous, and interception in preliminary stage. Three examples to demonstrate the general suitability of proposed method are presented in this section, including the Earth-Mars rendezvous; the Earth-Mars-Ceres rendezvous mission; and a hazardous asteroid interception mission. It shows that the proposed method can parameterize feasible trajectories using only two variables, so it is simpler to get the initial guess for continuous thrust trajectory. Secondly, compared with the shape-based method, it requires more fuel cost using the virtual gravity method, especially in the case of rendezvous problem with a long flight time. However, the capability of changing the speed and the position in radical direction in short time outperforms the shape-based method. This advantage can be used in rapid maneuver trajectory design, such as minimal time interception trajectory design for asteroid deflection missions.

2.3 Orbital Rendezvous Between Close Near Earth Asteroids Considering the Third Body Perturbation

2.3.1 Introduction

The Near Earth Asteroids missions provide significant additional return on the investments, including providing a large science return through in situ observation and returned sample, and carrying out an assessment of the resource available in the NEAs population that could be used for future space utilization. More than 500 close NEAs have low eccentricity orbits [22]. The source of these NEAs can be utilized for deep space exploration mission. The design of a multiple NEAs rendezvous mission therefore requires the assessment of a large number of orbital transfer trajectories [10]. It is prohibitively time consuming to evaluate all possible NEAs rendezvous tours using numerical integration methods. Therefore, approximate analytical methods that can rapidly estimate the fuel consumption of orbital rendezvous is required.

As the flight time of the low-thrust transfer is usually long compared to the impulsive transfer, the ignorance of the third-body perturbation can result in error. There are many literatures that study this problem. A patching conic method was proposed in [4], in which the orbital transfer in three-body dynamics is decomposed into two Keplerian approximations. The problem with that method is that, the error between the patching conic approximation method and the true solution of three-body problem is significant. In Ref. [17], the magnitude of the third-body attraction is assumed to be some orders lower than the gravity of the major central body, and the disturbing function of the third-body effect is expanded in power series. Recently, a semi-analytical approach to study the distance encounter between the spacecraft and the Earth was proposed in [3], in which the third-body perturbation for the Keplerian orbits is studied. It shows that the proposed method can keep a good accuracy compared to the numerical integration method. However, it can only be utilized to estimate the third-body perturbation for Keplerian orbits. In this section, the Earth gravitational perturbation is considered for the NEAs low-thrust rendezvous, and the change of orbital elements caused by which is estimated by the formulation proposed in [2]. Then, based on the virtual gravity field method proposed in Ref. [20], the variation of the orbital element caused by the low-thrust can be analytically solved. Therefore, given the states at the initial point and the terminal point, the NEAs orbital rendezvous problem can be converted to a problem of finding the parameters of the virtual gravity field. Unlike the shape-based method, there is no constraint about the direction of the thrust or the shape of the rendezvous transfer trajectory. Thus a large number of feasible trajectories that satisfy the boundary constraints can be obtained, and the optimal one can be calculated using the PSO algorithm.

In existing literatures, a large number of NEAs orbital rendezvous missions have been considered using the low thrust propulsion systems like the SEP or solar sailing [13, 16]. However, both of the solar sail and the SEP system have their limitations: the lifetime of the SEP mission is constrained by the amount of the propellant that the spacecraft can carry, whilst the solar sail concepts are constrained by the direction and the magnitude of the low thrust that the solar sail system can provide. Thus, the combination of a solar sail system with the SEP system is considered. The complementary nature of the solar sail and the SEP system enable various mission concepts. In these applications, the solar sail is utilized to generate the radial thrust to reduce the heliocentric gravity, and then the orbital rendezvous using the hybrid systems can be modeled as a low-thrust transfer in a reduced gravitational field, and the indirect optimization method is utilized to generate the optimal solution. In this section, a novel analytical model for the hybrid systems is developed for the NEAs orbital rendezvous. In the hybrid systems, the thrust generated by the solar sail is modeled as a function of the transfer angular, and then the thrust generated by the SEP can be expressed analytically. Therefore, the optimal minimal fuel consumption orbital transfer problem is converted to an optimization method, and the parameters of the virtual gravity field and coefficients of the attitude function are optimized using the PSO method to minimize the fuel consumption of the low-thrust NEAs orbital rendezvous.

2.3.2 *The Variation of Orbital Elements Caused by the Earth Gravitational Perturbation*

In the NEAs orbital rendezvous, the Earth gravitational potential function, from [2], can be approximately written as,

$$R \doteq \mu \left(\frac{1}{r} + \frac{\cos\theta}{r^2} - \frac{1}{\sqrt{1 + r^2 - 2r\cos\theta}} \right) \tag{2.17}$$

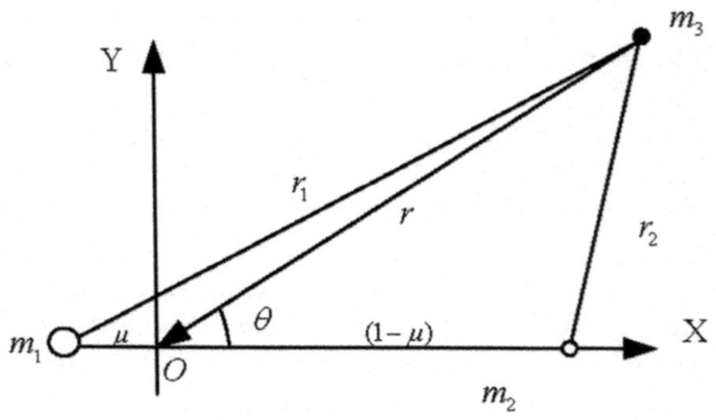

Fig. 2.16 The third body perturbation

where μ is the normalized mass of the the Earth, and the relative position of the spacecraft to the Earth is shown in Fig. 2.16, and the change of orbital elements caused by the Earth gravity potential function can be expressed using a semi-analytical formulation in [3],

Assuming that $\mu \ll 1$ $(\cos^2\theta) \approx 1$, it has,

$$r = \sqrt{{r_1}^2 - 2\mu r_1 \cos\theta + \mu^2} = r_1 - \mu\cos\theta + o(\mu^2) \tag{2.18}$$

while

$$\frac{r}{rr_1} = \frac{1}{r_1} = \frac{r_1}{rr_1} - \frac{\mu\cos\theta}{rr_1} + o(\mu^2) \approx \frac{1}{r} - \frac{\mu\cos\theta}{r^2} + o(\mu^2) \tag{2.19}$$

Thus the Hamilton Function can be written as,

$$H_{iner} = K + U + \mathrm{O}(\mu^2) \tag{2.20}$$

in which, the first term in Eq. 2.20 is the Keplerian term,

$$\mathrm{K} = \frac{1}{2}({p_x}^2 + {p_y}^2 + {p_z}^2) - \frac{1}{r} \tag{2.21}$$

The second term is the perturbation term, and it can be written as,

$$U = \mu\left(\frac{1}{r} + \frac{\cos\theta}{r^2} - \frac{1}{r_2}\right) \tag{2.22}$$

As

$$r = \sqrt{{r_1}^2 - 2\mu r_1\cos\theta + \mu^2} \tag{2.23}$$

Fig. 2.17 The relative position and the phasing angle between the Earth and the satellite

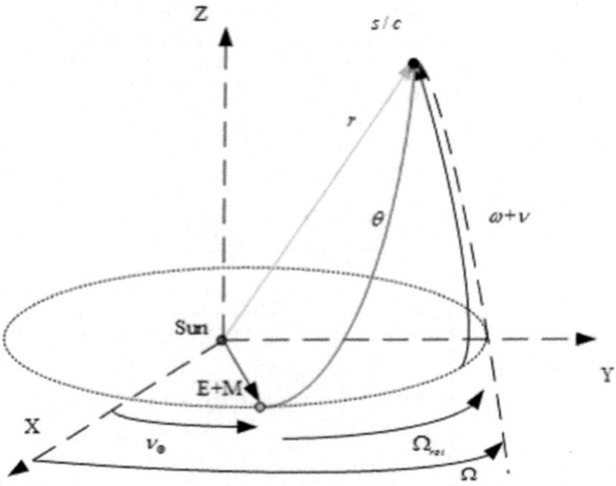

(a) the relative position of the satellite and the earth.

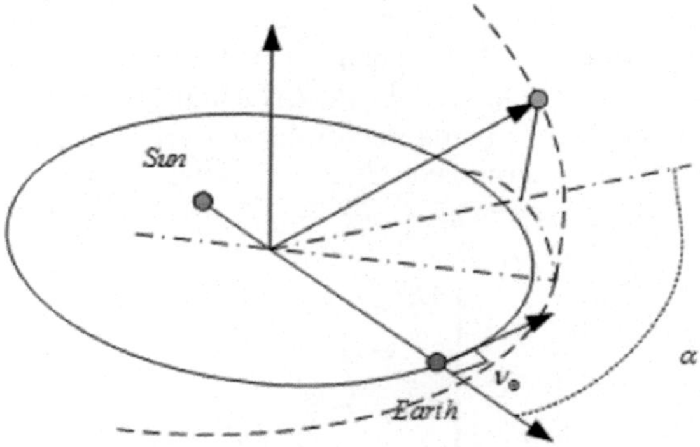

(b) the phasing angle between the Earth and the satellite.

Substitute Eq. 2.23 into Eq. 2.22, and it has,

$$U \approx -\mu\left(\frac{1}{r}+\frac{\cos\theta}{r}-\frac{1}{\sqrt{1+r^2-2r\cos\theta}}\right) \tag{2.24}$$

As shown in Eq. 2.24, the third body perturbation it determined by θ, r. In Fig. 2.17, it has,

$$\cos\theta = \cos\Omega_{rot}\cos(\omega+v)-\sin\Omega_{rot}\sin(\omega+v)\cos i \tag{2.25}$$

In the X-Y plane, if the spacecraft is in front of the Earth, it has $\Omega_{rot} = \Omega - v_{\Theta}$. The change of the orbital elements caused by the third body can be written as,

$$\begin{cases} \Delta a = \int\limits_{\theta_0}^{\theta_f} \frac{2}{n^2a^2} \frac{\partial R}{\partial M_0} \mathrm{d}v \\ \Delta \imath = -\left(\int\limits_{\theta_0}^{\theta_f} \frac{1}{n^2a^2\sqrt{1-e^2}\sin i} \frac{\partial R}{\partial \Omega} \mathrm{d}v - \int\limits_{\theta_0}^{\theta_f} \frac{\cos i}{n^2a^2\sqrt{1-e^2}\sin i} \frac{\partial R}{\partial \omega} \mathrm{d}v \right) \\ \Delta \Omega = \left(\int\limits_{\theta_0}^{\theta_f} \frac{1}{n^2a^2\sqrt{1-e^2}\sin i} \frac{\partial R}{\partial \imath} \mathrm{d}v \right) \\ \Delta \omega = \int\limits_{\theta_0}^{\theta_f} \frac{\sqrt{1-e^2}}{n^2a^2e} \frac{\partial R}{\partial e} \mathrm{d}v - \int\limits_{\theta_0}^{\theta_f} \frac{\cos i}{n^2a^2\sqrt{1-e^2}\sin i} \frac{\partial R}{\partial i} \mathrm{d}v \\ \Delta M = n - \int\limits_{\theta_0}^{\theta_f} \frac{2}{n^2a} \frac{\partial R}{\partial a} \mathrm{d}v - \int\limits_{\theta_0}^{\theta_f} \frac{1-e^2}{n^2a^2e} \frac{\partial R}{\partial e} \mathrm{d}v \\ \Delta e = \int\limits_{\theta_0}^{\theta_f} \frac{1-e^2}{n^2a^2e} \frac{\partial R}{\partial M_0} \mathrm{d}v - \int\limits_{\theta_0}^{\theta_f} \frac{\sqrt{1-e^2}}{n^2a^2e} \frac{\partial R}{\partial \omega} \mathrm{d}v \end{cases} \tag{2.26}$$

where $a, e, i, \omega, \Omega, M$ are the semi-major axis, the eccentricity, the inclination, the argument of periapsis, the RAAN and the eccentric anomaly, respectively; n is the nominal mean motion rate of the spacecraft; θ_0, θ_f are true anomaly of the initial orbit and the terminal orbit, respectively.

The expression of $\frac{\partial R}{\partial a}, \frac{\partial R}{\partial e}, \frac{\partial R}{\partial i}, \frac{\partial R}{\partial \Omega}, \frac{\partial R}{\partial \omega}, \frac{\partial R}{\partial v}$ in Eq. 2.26 can be written as,

$$\begin{cases} \frac{\partial R}{\partial a} = \mu A \frac{1-e^2}{1+e\cos(v)} - \mu B \frac{3M}{2n} C \\ \frac{\partial R}{\partial e} = \mu CA + \mu CB \\ \frac{\partial R}{\partial \Omega} = \mu BC \\ \frac{\partial R}{\partial i} = \mu B \sin(\Omega_{rot}) \sin(\omega + v) \cos(i) \\ \frac{\partial R}{\partial \omega} = \mu BD \\ \frac{\partial R}{\partial M} = \frac{a^2\sqrt{(1-e^2)}}{r^2} \frac{\partial R}{\partial v} \\ \frac{\partial R}{\partial v} = -\mu \frac{e\sin(v)}{r(1+e\cos(v))} - \frac{1}{r^2} D - \frac{2e\cos(\theta)\sin(v)}{r^2(1+e\cos(v))} - \\ \frac{\mu}{(1+r^2-2r\cos(\theta))^{3/2}} (r^2 \frac{e\sin(v)}{1+e\cos(v)} + rD) \\ + \frac{\mu r\cos(\theta)}{(1-r^2-2r\cos(\theta))} \frac{e\sin(v)}{1+e\cos(v)} \end{cases} \tag{2.27}$$

where

$$\begin{cases} A = -\frac{1}{r^2} - 2\frac{\cos(\theta)}{r^3} + \frac{(r-\cos(\theta))}{(1+r^2-2r\cos(\theta))^{3/2}} \\ B = \frac{1}{r^2} - \frac{r}{(1+2r^2-2r\cos(\theta))^{3/2}} \\ C = \sin(\Omega + v_e)\cos(\omega + v) + \cos(\Omega + v_e)\sin(\omega + v)\cos(i) \\ D = \cos(\Omega + v_e) + \sin(\Omega + v_e) cos(\omega + v) \end{cases} \tag{2.28}$$

where v_e is the true anomaly of the Earth. The relative angular distance α_0, can be defined as the angle between the Sun-Earth line and the projection of the periapsis line of the spacecraft on to the Earth orbital plane, at the moment of the periapsis passage by t_0. The phasing angular difference can be computed as,

$$\alpha = \Omega(t_0) - v_e(t_0) + \tan(\cos(i_0 \tan(\omega_0))) \tag{2.29}$$

2.3.3 Orbital Rendezvous Considering the Third Body Perturbation

For the three-dimensional orbital transfer, the change of orbital elements can be divided into co-planer change (Δa, Δe, $\Delta\omega$, ΔE) and out-of-plane change (Δi, $\Delta\Omega$). Under the constant radial thrust A_R (outward) and constant circumferential thrust A_θ, the differential equation of the orbital elements can be written as,

$$\begin{cases} \frac{da}{dE} = \frac{2a^3}{\mu}(A_R e\sin(E) + A_\theta\sqrt{(}1-e^2)) \\ \frac{de}{dE} = \frac{a^2}{\mu}[A_R(1-e^2)\sin(E) + A_\theta(2\cos(E) - e - e(\cos(E))^2)\sqrt{(}1-e^2)] \\ \frac{d\omega}{dE} = -\frac{a^2}{e\mu}[A_R(\cos(E)-e)\sqrt{(}1-e^2) + A_\theta(2-e^2-e\cos(E))\sin(E)) \\ \frac{dE}{dt} = \sqrt{\mu/a^3} \end{cases} \tag{2.30}$$

where a, e, ω and are the semi-major axis, the eccentricity, the argument of the periapsis and the eccentric anomaly, respectively. The change of these orbital elements can be approximately obtained by integrating Eq. 2.30 from the initial point (E_i) to the terminal point (E_f), as show in Eq. 2.31,

$$\begin{cases} \Delta a = C_1 A_R + C_2 A_\theta \\ \Delta e = C_3 A_R + C_4 A_\theta \\ \Delta\omega = C_5 A_R + C_6 A_\theta \end{cases} \tag{2.31}$$

where $C_1 \sim C_6$ are

$$\begin{cases} C_1 = -\frac{2a^3}{\mu}e(\cos E_f - \cos E_i) \\ C_2 = \frac{2a^3}{\mu}\sqrt{1-e^2}(E_f - E_i) \\ C_3 = -\frac{a^2}{\mu}(1-e^2)(\cos E_f - \cos E_i) \\ C_4 = \frac{a^2}{\mu}\sqrt{1-e^2}(2\cdot(\sin E_f - \sin E_i) - \frac{e}{4}(\sin 2E_f - \sin 2E_i) - \frac{e}{2}\cdot(E_f - E_i)) \\ C_5 = -\frac{a^2}{e\mu}\sqrt{1-e^2}((\sin E_f - \sin E_i) - e\cdot(E_f - E_i)) \\ C_6 = -\frac{a^2(2-e^2)}{e\mu}\cdot(\cos 2E_f - \cos 2E_i) + \frac{a^2}{2\mu}(\cos 2E_f - \cos 2E_i) \end{cases} \tag{2.32}$$

The change of the inclination and the right ascension of the ascending node (RAAN) are $\Delta i = i_t - i_i$ and $\Delta\Omega = \Omega_t - \Omega_i$. For the case of the orbital rendezvous trajectory, the seventh-degree polynomials function are applied for orbital transfer, as shown in Eq. 2.33,

$$\begin{cases} \Delta i = c_0 + c_1\theta + c_2\theta^2 + c_3\theta^3 + c_4\theta^4 + c_5\theta^5 + c_6\theta^6 + c_7\theta^7 \\ \Delta\Omega = d_0 + d_1\theta + d_2\theta^2 + d_3\theta^3 + d_4\theta^4 + d_5\theta^5 + d_6\theta^6 + d_7\theta^7 \end{cases} \tag{2.33}$$

the boundary constraints are,

$$\begin{cases} \Delta i(\theta_0) = \Delta\Omega(\theta_0) = 0,\ \Delta i(\theta_f) = i_t - i_i;\ \Delta\Omega(\theta_f) = \Omega_t - \Omega_i \\ \left.\frac{\partial^j \Delta i}{\partial\theta^j}\right|_{\theta_0} = \left.\frac{\partial^j \Delta\Omega}{\partial\theta^j}\right|_{\theta_0} = 0,\ \left.\frac{\partial^j \Delta i}{\partial\theta^j}\right|_{\theta_f} = \left.\frac{\partial^j \Delta\Omega}{\partial\theta^j}\right|_{\theta_f} = 0 \end{cases} \tag{2.34}$$

thus the partial coefficients in Eq. 2.33 are $c_0 = c_1 = c_2 = c_3 = d_0 = d_1 = d_2 = d_3 = 0$, and $c_i, d_i (i = 4, 5, 6, 7)$ can be calculated using undetermined coefficient method. Then through Eq. 2.34, the inclination and RAAN Ω of the transfer trajectory can be obtained. Assuming the elevation angle is φ, and then from Ref. [2], it has,

$$\begin{aligned} \cos\varphi = \cos(\varphi(i, \Omega)) &= \cos(\theta_0 + \omega_i)\cos(\theta + \omega)\cos(\Omega - \Omega_i) + \cos i_i \sin(\theta_0 + \omega_i)\cos(\theta + \omega) \\ &- \cos i \cos(\theta_0 + \omega_i)\sin(\theta + \omega)\sin(\Omega - \Omega_i) \\ &+ \sin(\theta_0 + \omega_i)\sin(\theta + \omega)[\sin i_i \sin i + \cos i_i \cos i \cos(\Omega - \Omega_i)] \end{aligned} \tag{2.35}$$

Then the required out of plane thrust, from [26], can be expressed as,

$$T_n = \frac{r}{\sqrt{(\varphi')^2 + \cos^2\varphi}}(\cos\varphi(\varphi'' - \sin\varphi\cos\varphi) + 2\sin\varphi((\varphi')^2 + \cos^2\varphi))) \tag{2.36}$$

the expression of φ' φ'' can be derived from Eq. 2.27. The fundamental work of trajectory design is to find the thrust that can propel the spacecraft from the initial point to the target point. Here the boundary constraints are expressed in terms of orbital elements. Assuming the orbital elements of the initial orbit are $a_i, e_i, i_i, \Omega_i, \omega_i, \theta_i$ and the orbital elements of target orbit are $a_t, e_t, i_t, \Omega_t, \omega_t, \theta_t$, the differences of the orbital elements are,

$$\Delta a = a_t - a_i;\ \Delta e = e_t - e_i;\ \Delta i = i_t - i_i;\ \Delta\Omega = \Omega_t - \Omega_(i);\ \Delta\omega = \omega_t - \omega i. \tag{2.37}$$

The difference in terms of the orbital elements consists of two parts, the first one is caused by the Earth gravity perturbation, and the second one is caused by the low-thrust. Thus the boundary constraints can be written as,

$$\begin{cases} \Delta a = \Delta a_{2b} + \Delta a_{3bp} \\ \Delta e = \Delta e_{2b} + \Delta e_{3bp} \\ \Delta\omega = \Delta\omega_{2b} + \Delta\omega_{3bp} \\ \Delta i = \Delta i_{2b} + \Delta i_{3bp} \\ \Delta\Omega = \Delta\Omega_{2b} + \Delta\Omega_{3bp} \end{cases} \tag{2.38}$$

where Δa_{2b}, Δe_{2b}, $\Delta\omega_{2b}$, Δi_{2b}, $\Delta\Omega_{2b}$ are the changes of orbital elements caused by the low-thrust, and Δa_{3bp}, Δe_{3bp}, $\Delta\omega_{3bp}$, Δi_{3bp}, $\Delta\Omega_{3bp}$ are the change of orbital elements caused by the Earth gravity perturbation. As the Earth gravitational perturbation can be calculated by the semi-analytical formulation of Eq. 2.26 it is required to find the feasible low thrust to satisfy the boundary constraints of Eq. 2.38.

Based on the virtual gravity filed method, the semi-analytical solutions for the low-thrust transfer considering the Earth gravity perturbation can be obtained. First, a virtual gravity field (μ_{vg},$\boldsymbol{r}_{vg}$) is given. Secondly, given the initial states and the terminal states of the spacecraft in the virtual gravity field, the initial orbital elements a_{vg_1}, e_{vg_1}, ω_{vg_1} and the target orbital elements a_{vg_2}, e_{vg_2}, ω_{vg_2} in the virtual gravity field can be obtained. Thirdly, given the initial orbital elements, the change of the orbital elements caused by the Earth gravitational perturbation can be calculated using Eq. 2.26. And the change of orbital elements caused by the low-thrust propulsion can be obtained as,

$$\begin{cases} \Delta a_{2b} = a_{vg\_2} - a_{vg\_1} - \Delta a_{3bp} \\ \Delta e_{2b} = e_{vg\_2} - e_{vg\_1} - \Delta e_{3bp} \\ \Delta\omega_{2b} = \omega_{vg\_2} - \omega_{vg\_1} - \Delta\omega_{3bp} \end{cases} \tag{2.39}$$

The spacecraft that propelled by the low-thrust is assumed to be transferred from the initial point to the target point in a virtual gravity field. For the close NEAs with small eccentricity, $C_1 \approx 0$, and then Eq. 2.39 can be written as,

$$\begin{cases} \Delta a_{2b} = C_2 A_\theta \\ \Delta e_{2b} = C_3 A_\theta + C_4 A_R \\ \Delta\omega_{2b} = C_5 A_\theta + C_6 A_R \end{cases} \tag{2.40}$$

Given the orbital elements of the initial and target points in the virtual gravity frame, the coefficients of C_2, C_3, C_4, C_5, C_6 can be calculated through Eq. 2.32 in appendix, and the required radial thrust and circumferential thrust can be calculated using Eq. 2.40 as,

$$\begin{cases} A_\theta = \frac{\Delta a_{2b}}{C_2} \\ A_R = \frac{\Delta e_{2b} - C_4 A_c}{C_3} \end{cases} \tag{2.41}$$

Substitute Eq. 2.41 into Eq. 2.40, the error of the argument of the periapsis can be calculated as

$$\varepsilon_\omega = \omega_{vg_2} - \omega_{vg_1} - \Delta\omega_{3bp} - (C_5 A_\theta + C_6 A_R) \tag{2.42}$$

The feasible virtual gravity parameters should satisfy the boundary constraint $\varepsilon_\omega < \omega^*$, where is the tolerance error. As shown in Fig. 2.18, after obtaining the parameter of the virtual gravity field, the required radial thrust and circumferential thrust in the heliocentric gravity field, can be calculated as,

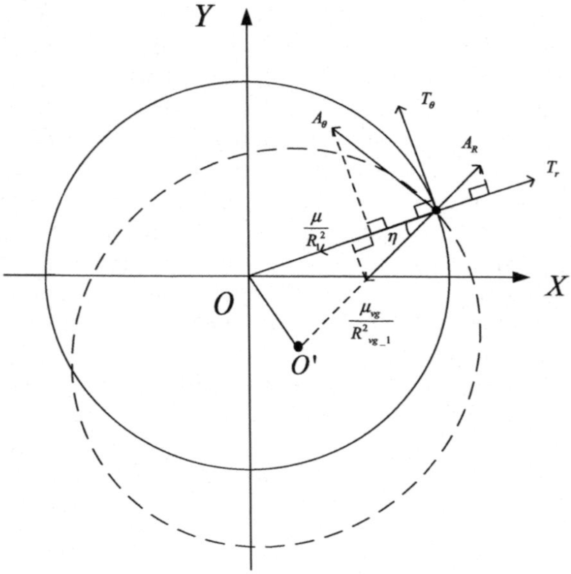

Fig. 2.18 Thrust required for orbital rendezvous

$$\begin{cases} T_r = \frac{\mu}{R_1{}^2} - \frac{\mu_{vg}}{R_{vg\_1}{}^2}\cos(\eta) + A_R\cos(\eta) - A_\theta\sin(\eta) \\ T_\theta = -\frac{\mu_{vg}}{R_{vg\_1}{}^2}\sin(\eta) + A_\theta\cos(\eta) + A_R\sin(\eta) \end{cases} \tag{2.43}$$

For the out-of-plane transfer case, T_n can be calculated using Eq. 2.36, and the orbital rendezvous considering the Earth gravitational perturbation can be solved by searching the parameters of the virtual gravity field to satisfy the boundary constraints of Eq. 2.37.

2.3.4 *Minimal Fuel Consumption Optimization Using Hybrid Systems*

2.3.4.1 The Model of the Hybrid Low-Thrust Systems

For the hybrid systems, the total required thrust can be written as,

$$\boldsymbol{T}_{total} = \boldsymbol{T}_{ss} + \boldsymbol{T}_{SEP} \tag{2.44}$$

where $\boldsymbol{T}_{ss} = [f_{ssr}, f_{ss\theta}]$ is the thrust generated by the solar sail and $\boldsymbol{T}_{SEP} = [f_{SE\Pr}, f_{SEP\theta}, f_{SEPh}]$ is the thrust generated by the SEP system. The solar sail system model in [14] is utilized, and the mathematic model for the solar sail is,

$$\begin{cases} f_{ssr} = \frac{\beta_\sigma \mu}{2r^2} \cos\alpha_{ss} \cdot (b_1 + b_2 \cos^2\alpha_{ss} + b_3 \cos\alpha_{ss}) \\ f_{ss\theta} = \frac{\beta_\sigma \mu}{2r^2} \sin\alpha_{ss} \cdot \cos\alpha_{ss} \cdot (b_2 \cos\alpha_{ss} + b_3) \end{cases} \tag{2.45}$$

where α_{ss} means the cone angle of the solar sail panel, β_σ is the solar sail lightness number, and b_1, b_2, b_3 are the coefficients related to the optical properties of the reflective film. The SEP system in [14] is applied here, in which the propulsion system can adjust both the magnitude and the direction of the thrust. In the hybrid systems proposed in Ref. [14], the solar sail is utilized to generate the radial thrust, and the SEP is utilized to generate the circumferential thrust. Be different to the hybrid system in [14], the hybrid systems proposed in this paper is assumed that, the solar sail is utilized to generate both the radial thrust and the circumferential thrust, and the SEP is used to produce the remainder thrust. In the hybrid systems, the required thrust can expressed as,

$$\begin{cases} f_{reqr} = f_{ssr} + f_{SE\,\mathrm{Pr}} \\ f_{req\theta} = f_{ss\theta} + f_{SEP\theta} \\ f_{reqh} = f_{SEPh} \end{cases} \tag{2.46}$$

As shown in Eq. 2.46, the thrust produced by the solar sail and the SEP is coupled. Here the cone angle of the solar sail panel is modeled as a function of transfer angle $\varDelta E$,

$$\alpha_{ss} = \pi \cos(l_2 \cdot \varDelta E^2 + l_1 \cdot \varDelta E + l_0) \tag{2.47}$$

where $l_i (i = 0, 1, 2)$ are coefficients of the cone angle function, $l_0 \in [0, 2\pi]$, $l_1 \in [0, 1]$, $l_2 \in [0, 1]$. Given the required thrust generated by the hybrid systems, the required thrust produced by the SEP is determined by the coefficients of solar sail's cone angle, as shown in Eq. 2.48,

$$\begin{cases} f_{SE\,\mathrm{Pr}} = f_{reqr} - f_{ss\theta} = F_r(l_0, l_1, l_2) \\ f_{SEP\theta} = f_{req\theta} - f_{ss\theta} = F_t(l_0, l_1, l_2) \end{cases} \tag{2.48}$$

2.3.4.2 Minimal Fuel Consumption of the NEAs Rendezvous Optimization Using Hybrid Systems

As the NEAs low-thrust rendezvous trajectory can be parameterized using the variable set $(\mu_{vg}, \boldsymbol{r}_{vg})$, and the required thrust of the hybrid systems can be parameterized using the coefficients (l_0, l_1, l_2). The fuel cost of the NEAs orbital rendezvous can be parameterized using $(\mu_{vg}, \boldsymbol{r}_{vg}, l_0, l_1, l_2)$. Therefore, the minimal fuel consumption of the low-thrust NEAs rendezvous trajectory optimization problem addressed here is to find the optimal parameters $X = (\mu_{vg}, \boldsymbol{r}_{vg}, l_0, l_1, l_2)$ to minimize the velocity increment, which is defined as,

$$J = \Delta V = F_{fuel}(\mu_{vg}, \boldsymbol{r}_{vg}, l_0, l_1, l_2) \tag{2.49}$$

Considering the magnitude of the low thrust that the solar sail and the SEP can provide, the constraints of the free variables are $\eta = \mu_{vg}/\mu_0 \in [0.8, 1.2]$, $l_0 \in [0, 2\pi]$, $l_1 \in [0, 1]$, $l_2 \in [0, 1]$. Here a two-level PSO algorithm is applied to solve this optimization problem. It is assumed that the iteration number for PSO level one is M_1, and for PSO level two is M_2. The procedure of the PSO algorithm is illustrated as follows,

Step 1: Provide the initial state and the terminal state of the spacecraft; then give the range of parameters $\mu_{vg} \in [\mu_{vgL}, \mu_{vgU}]$, $r_{vgx} \in [r_{vgxL}, r_{vgxU}]$, $r_{vgy} \in [r_{vgyL}, r_{vgyU}]$.

Step 2: Initialize the particle? position with a uniformly distributed random vector $P_i = (\boldsymbol{r}_{vg}, \mu_{vg})$.

Step 3: Calculate the initial state and the terminal state in the virtual gravity field using Eq. 2.1.

Step 4: Estimate the change of orbital elements caused by the Earth gravity perturbation using Eq. 2.26; then calculate the change of orbital elements caused by the low-thrust using Eq. 2.39.

Step 5: Calculate the corresponding transfer trajectory parameterized using variable set $\boldsymbol{r}_{vg}, \mu_{vg}$); then calculate the error using Eq. 2.42. Here the boundary constraint tolerance error is set, if the error satisfies the boundary constraint, turn to step6, else turn to step2.

Step 6: Calculate the required low-thrust using Eq. 2.43.

Step 7: Initialize the particle? position with a uniformly distributed random vector $G_i = (l_0, l_1, l_2)_j$.

Step 8: Substitute $(l_0, l_1, l_2)_j$ into Eq. 2.48, and calculate the fuel consumption of the hybrid systems F_j, $i \in [1, M_1]$ using Eq. 2.49. Initialize the particle best-known position to its initial position; then update the swarm? best known position and its fitness. If the number of iteration$j < M_1$, turn to step7, if number of iteration $j > M_1$, turn to step9.

Step 9: If $i < M_2$, turn to step2, else turn to step 10.

Step10: Calculate the minimal value $J = \min\{F_i\}, 0 < i < M_2$.

2.3.5 Simulation Analysis

In order to testify the validity of the proposed method, the NEAs tour mission in [10] is applied is this simulation section. The initial orbital elements of the NEAs are shown in Table 2.8, and three cases of NEAs rendezvous are utilized in this simulation: case 1, orbital rendezvous from 2000SG344 to 2015 JD 3; case 2, orbital rendezvous from 2015 JD 3 to 2012 KB4; case 3, orbital rendezvous from 2012 KB4 to 2008 EV5.

The Earth gravitational perturbations for the low-thrust rendezvous between the NEAs are estimated. Considering the magnitude of the low-thrust that the hybrid

Table 2.8 Orbit elements of the NEA and Earth (MJD2000:4748.5)

Planet	a/AU	e	i/deg	ω/deg	Ω/deg	M/deg
2000 SG344	0.978	0.0669	0.111	275.167	192.072	359.944
2015 JD 3	1.058	0.0082	2.722	63.609	43.875	165.158
2012 KB4	1.093	0.061	6.328	277.683	70.498	63.825
2008 EV5	0.958	0.084	7.437	234.785	93.398	193.073

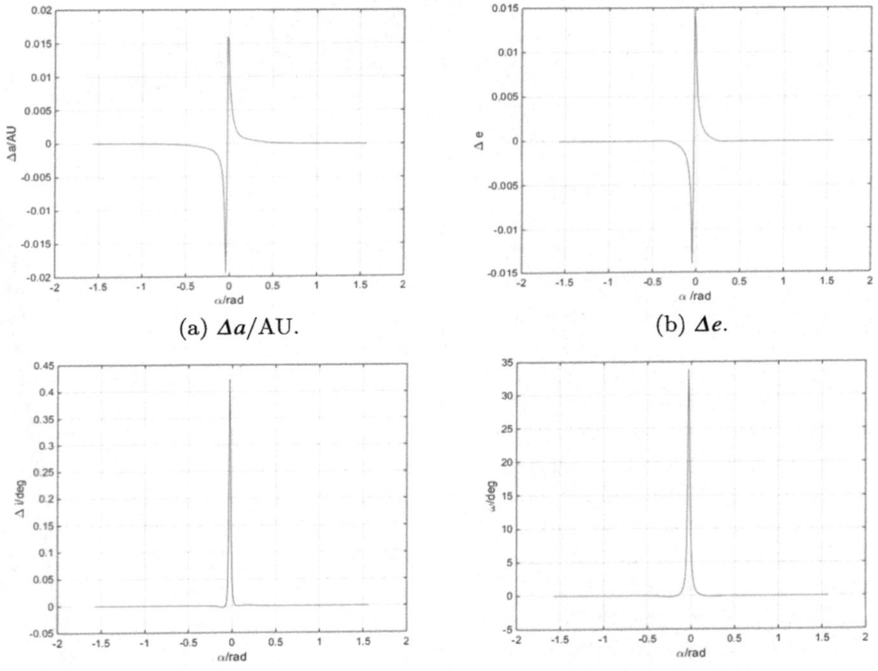

(a) Δa/AU. (b) Δe.

(c) Δi/deg. (d) ω/deg.

Fig. 2.19 The Earth gravity perturbation in terms of Δa,Δe,Δi,$\Delta\omega$

low-thrust systems can provide, the constraints of the parameters of the virtual gravity field are set as $\eta = \mu_{vg}/\mu_0 \in [0.8, 1.2]$, $r_{vgx} \in [0, 0.2]AU$, $r_{vgy} \in [0, 0.2]AU$, and the constraints of coefficients of the attitude function for the solar sail are $l_0 \in [0, 2\pi]$,$l_1 \in [0, 1]$,$l_2 \in [0, 1]$ The tolerance error is set as $\varepsilon^* = 10^{-4}$. In the two-level PSO algorithm, the iterations number is set as $M_1 = M_2 = 1000$. The parameters of the virtual gravity field and the coefficients of the attitude are optimized to minimize the fuel consumption of the NEAs orbital rendezvous.

The Earth gravitational perturbation for the low-thrust NEAs orbital rendezvous can be calculated using Eq. 2.26. As the perturbation is mainly determined by the phasing difference between the Earth and the spacecraft, a slight change of the orbital

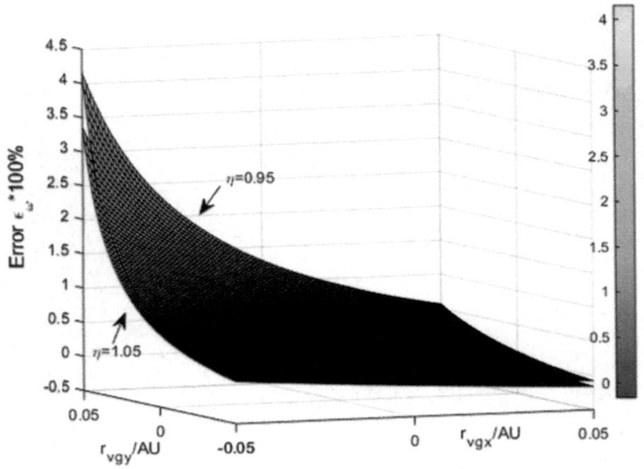

Fig. 2.20 The boundary constraints error

elements caused by the low-thrust can rarely affect the perturbations. Therefore the semi-analytical formulation of Eq. 2.26 can estimate the third body perturbation accurately. In this case, taking the example of orbital rendezvous from the NEA 2003SG 34 to the NEA 2015JD, the range of the phasing angular difference is set as $\alpha \in [-\pi/2, \pi/2]$ rad, and the transfer angular is set as $\theta = 2\pi$. Then the change of the orbital elements in terms of a, e, i, ω caused by the Earth gravitational perturbation is shown in Fig. 2.19. It shows that in case that the phasing angle difference between the Earth and the spacecraft (α) is large, the perturbation is very small and can be ignored. Otherwise, if the phasing difference is small (like $\alpha \in [-0.5, 0.5]$ rad), the perturbation is obvious and cannot be ignored. The formulation of Eq. 2.26 is utilized to estimate the change of orbital elements for these three cases. For case1 and case2, the phasing angle difference can be calculated. It can obtain that $\alpha_0 > 0.5$ rad, thus the Earth gravitational perturbation is very small and can be ignored. For case 3, the changes of orbital elements caused by the Earth gravitational perturbation in terms of orbital elements are $\Delta a_{3bp} = 2.71 \times 10^{-3}$AU. Although the changes are small in this case, the consideration of the Earth gravitational perturbation can improve the accuracy for precise low-thrust rendezvous.

It should be pointed out that, compared to the shape based method the advantage of the proposed method is that it can generate a large number of feasible rendezvous transfers that can satisfy the boundary constraints. The main reason is that there is no constraint on the direction of the low thrust or the shape of the rendezvous trajectory. Taking case1 as an example, the magnitude of the virtual gravity field is set as $\mu_{vg} = 0.95\mu_0$ and $\mu_{vg} = 1.05\mu_0$, then the position vector of the virtual gravity field are searched to satisfy the boundary constraints. The errors are shown in Fig. 2.20. It can be found that a large number of parameters set $(\boldsymbol{r}_{vg}, \mu_{vg})$, which satisfy $\varepsilon < \varepsilon^*$, can be found. Thus there are a large number of corresponding feasible rendezvous trajectories that can satisfy the boundary constraints. The minimal fuel consumption of the orbital rendezvous trajectory can be obtained using the two-level PSO algorithm. The optimal parameters are shown in Table 2.9.

Table 2.9 Optimal parameters of the VCGF in each case

Transfer case	Case 1	Case 2	Case 3
$\boldsymbol{r}_{vg} = (\boldsymbol{r}_{vgx}, \boldsymbol{r}_{vgy})$	$\boldsymbol{r}_{vg}$ = (0.01,−0.00012)	$\boldsymbol{r}_{vg}$ = (0.065,0)	$\boldsymbol{r}_{vg}$ = (−0.0275,0.031)
$\eta = (-\mu_{vgx}/\mu_0)$	0.875	0.907	1.012

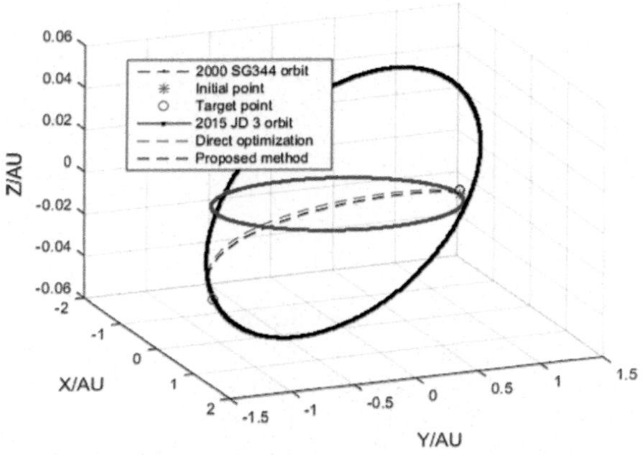

Fig. 2.21 2000SG344-2015 JD 3 orbital transfer (case1)

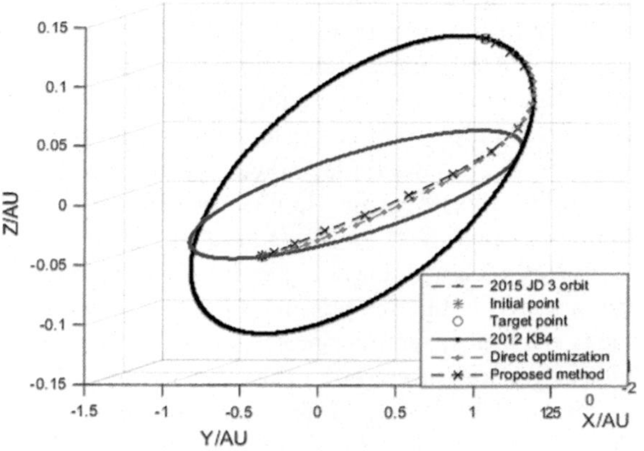

Fig. 2.22 2015 JD 3-2012 KB4 orbital transfer (case2)

Then the NEAs rendezvous using the proposed method is compared to the direct true optimization method. The solution generated by the proposed method is provided as the initial guess, and the true optimization solution is obtained using the numerical optimization method. The rendezvous transfer trajectories of three cases are illustrated as Figs. 2.21, 2.22 and 2.23. The speed increment of transfer trajectories using the proposed method is compared to that obtained using the numerical method, and the results are: 1). Case1 the fuel cost using the direct optimization method is $\Delta V_{op} = 37.12\,\mathrm{m/s}$, and the computational time is $t = 12.6\,\mathrm{s}$. The fuel cost using the proposed method is $\Delta V_{vg} = 39.10\,\mathrm{m/s}$ and the computational time $t = 4.4\,\mathrm{s}$; 2).

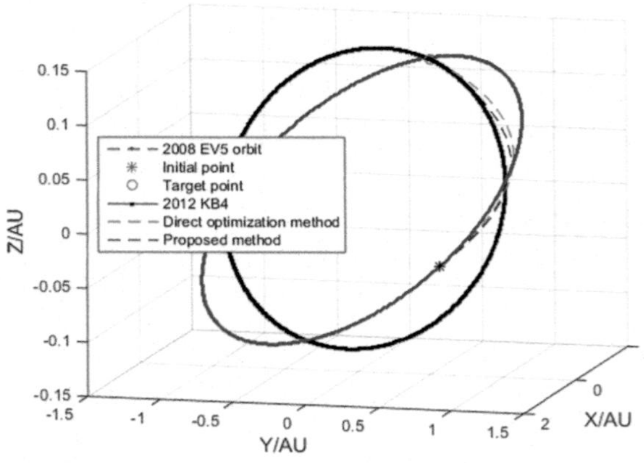

Fig. 2.23 2012 KB4-2008 EV5 orbital transfer (case3)

Case 2, the fuel cost using the direct optimization method is $\Delta V_{vg} = 57.732\,\text{m/s}$, and the computational time is $t = 18.1\,\text{s}$; the fuel cost using the proposed method is $\Delta V_{vg} = 61.551\,\text{m/s}$, and the computational time is $t = 5.72\,\text{s}$. 3).Case 3: the fuel cost using the direct optimization method is $\Delta V_{op} = 50.12\,\text{m/s}$ and the computational time is $t = 28.31\,\text{s}$; the fuel cost using the proposed method is $\Delta V_{vg} = 54.90\,\text{m/s}$, and the computational time is $t = 17.24\,\text{s}$. It can find that, the required speed increment for the low-thrust transfer of the proposed is slightly larger than the true optimal transfer, but the computation time required for the proposed method is much less than that using the true optimization solution. Thus one conclusion is that the proposed method can rapidly to generate the near optimal solution of the NEAs orbital rendezvous, the proposed method can serve an efficient tool for the preliminary stage, especially when there are a large number of feasible target orbital rendezvous needed to be evaluated.

The efficiency of the analytical method for hybrid systems proposed in this paper is compared to the one presented in [14], in which the solar sail is to generate only the radial thrust, and the circumferential thrust is provided by the SEP system. In the hybrid systems proposed in this section, the solar sail is used to generate both the radial thrust and the circumferential thrust, and the SEP system is utilized to generate the remainder thrust. The simulation results are already illustrated in Figs. 2.21, 2.22 and 2.23. For the orbital rendezvous case 1, the fuel cost of the thrust accounts for 63.33% of that using the traditional hybrid systems. For the orbital rendezvous case 2, the fuel cost accounts for 58.8% of that using the traditional one. For case 3, the result using the proposed hybrid systems is about 83.3% of that using the traditional one. It should be pointed out that, in orbital rendezvous case 3, when the transfer time reaches 16 days, the thrust generated by the solar sail is larger than the required thrust, and the SEP system generated the negative circumferential thrust. The main reason is that the thrust generated by the hybrid systems is constrained by Eq. 2.46. Although the radial thrust generated by the solar sail is slightly larger than the required thrust, it generates larger circumferential thrust, thus the thrust generated by the SEP is reduced, and the minimal fuel cost orbital rendezvous can

Table 2.10 Optimal parameters of the three transfer trajectories

Transfer case	Case 1	Case 2	Case 3
l_0	0	0	$(0,\pi/2)$
l_1	−0.2012	−0.0143	(−0.1962,0)
l_2	0.03651	0.0392	(0.0472,0)

be reduced, same phenomenon also exists in Ref. [14]. It should be pointed out that, in case of the orbital rendezvous case 3, the required radial thrust for the NEAs orbital rendezvous is negative when the transfer time reaches 58 days. But because the solar sail cannot generate negative thrust (direct to the Sun), thus it is set to be zero ($l_0 = \pi/2, l_1 = l_2 = 0$) after the transfer time reaches 58 days, as shown in Table 2.10. In this case, the consumption using the proposed method is about 83.3% of the one obtained using the traditional hybrid systems. Therefore, one conclusion can be obtained that the proposed hybrid systems has a higher efficiency compared to the traditional one.

2.3.6 Conclusion

A novel approach for the hybrid low-thrust propulsion orbital rendezvous between close NEAs considering the Earth gravitational perturbation is presented. Based on the virtual gravity field method, the semi-analytical solutions of the NEAs rendezvous transfer considering the Earth perturbation can be obtained. Then, in the hybrid systems, the required thrust generated by the hybrid system can be expressed analytically as a function by the coefficients of the solar sail attitude and the parameters of the virtual gravity field. Therefore, the minimal fuel consumption rendezvous transfer optimization problem is converted to a parameter optimization problem, and a two-level PSO algorithm is utilized to solve it. The proposed method is applied in a few NEAs orbital rendezvous cases, and a few conclusions can be obtained, 1) Since there is no constraint on the shape of the trajectory or the direction of the thrust in the proposed method, a large number of feasible semi-analytical solutions of the NEAs orbital rendezvous can be obtained. 2). Earth's gravitational perturbation is considered for the NEAs rendezvous, which can improve the accuracy of the orbital rendezvous especially when the phasing difference between the Earth and the probe is small, and the Earth perturbation cannot be ignored. 3). the solution of the fuel cost of hybrid systems can be obtained using the analytical hybrid systems model, and the near optimal one can be obtained quickly using the PSO algorithm. The proposed method can served as an efficient tool to estimate the fuel cost of close NEAs orbital rendezvous missions using hybrid systems at preliminary stage.

References

1. O. Abdelkhalik, E. Taheri, P.Y. Cui, D. Qiao et al., Approximate on-off low-thrust space trajectories using Fourier series. J. Spacecr. Rocket. **49**(5), 962–965 (2012)
2. E.M. Alessi, J.P. Sanchez Cuartielles, MOID-increasing disposal strategies for LPO missions, in *International Astronautical Congress*, IAC-14. C1 8-12 (2014)
3. E.M. Alessi, J.P. Sanchez, Semi-analytical approach for distant encounters in the spatial circular restricted three-body problem. J. Guid. Control **39**(2), 351–359 (2015)
4. R.R. Bate, D.D. Mueller, J.E. White, Fundamentals of astrodynamics. Acta Futur. **8**, 359–379 (1971)
5. J.T. Betts, Survey of numerical methods for trajectory optimization. J. Guid. Control Dyn. **21**(2), 193–207 (1998)
6. F.W. Boltz, Orbital motion under continuous radial thrust. J. Guid. Control Dyn. **14**(3), 667–670 (1991)
7. F.W. Boltz, Orbital motion under continuous tangential thrust. J. Guid. Control Dyn. **15**(6), 1503–1507 (1992)
8. Y. Gao, C.A. Kluever, Analytic orbital averaging technique for computing tangential-thrust trajectories. J. Guid. Control Dyn. **28**(6), 1320–1323 (2005)
9. A.L. Herman, B.A. Conway, Direct optimization using collocation based on high-order Gauss-Lobatto quadrature rules. J. Guid. Control Dyn. **19**(3), 592–599 (1996)
10. F. Jiang, Y. Chen, Y. Liu, Baoyin, et al. GTOC5: Results from the Tsinghua University, Gravity Assist Trajectories. Acta Futura **8**, 37–44 (2004)
11. J.A. Sims, P.A. Finlayson, Implementation of a Low-Thrust Trajectory Optimization Algorithm for Preliminary Design. *AIAA/AAS Astrodynamics Specialist Conference and Exhibit, Keystone, Colorado*, AIAA 2006-6746
12. S. Li, Y.S. Zhu, Y.K. Wang, Rapid design and optimization of low-thrust rendezvous/interception trajectory for asteroid deflection missions. Adv. Space Res. **53**, 696–707 (2014)
13. R.J. Mckay, M. Macdonald, J. Biggs et al., Survey of highly non Keplerian orbits with low thrust propulsion. J. Guid. Control Dyn. **34**(3), 645–666 (2011)
14. G. Mengali, A.A. Quarta, Trajectory design with hybrid low-thrust propulsion system. J. Guid. Control Dyn. **30**(2), 419–426 (2007)
15. G. Mengali, A.A. Quarta, Escape from elliptic orbit using constant radical thrust. J. Guid. Dyn. **32**, 1018–1022 (2009)
16. G. Mengali, A.A. Quarta, C. McInnes, Rapid solar sail rendezvous missions to asteroid 99942 Apophis. J. Spacecr. Rocket. **46**(1), 134–140 (2009)
17. O.P. Murioa, D.J. Scheeres, A perturbation theory. Acta Astronaut. **67**, 27–37, 2010
18. A.E. Petropoulos, J.M. Longuski, Shape-based algorithm for the automated design of low-thrust, gravity assist trajectories. J. Spacecr. Rocket. **41**(5), 787–796 (2004)
19. I. Shevchenko, The Kepler map in the three-body problem. New Astron. **16**(2), 94–99 (2011)
20. C. Sun, J.-P. Yuan, Q. Fang, Continuous low thrust trajectory optimization for preliminary design. Proc. Inst. Mech. Eng. Part G: J. Aerosp. Eng. **230**(5), 1–11 (2015)
21. H.S. Tsien, Take-off from satellite orbit. J. Am. Rocket Soc. **23**(4), 233–236 (1953)
22. G.B. Valsecchi, D. Vokrouhlicky, A. Milani, Near Earth objects, our celestial neighbors: opportunity and risk, in *Proceedings of the International Astronomical Union*, vol. 236, no. 236 (2007)
23. B.J. Wall, J.M. Longuski, Shape-based approximation method for low-thrust trajectory optimization. J. Astronaut., *AIAA/AAS Astrodynamics Specialist Conference and Exhibit* (Honolulu, USA, 2008)
24. B.J. Wall, B.A. Conway, Shape-based approach to low-thrust rendezvous trajectory design. J. Guid. Control. Dyn. **32**(1), 95–101 (2009)
25. C.H. Zee, Low tangential thrust trajectories improved first-order solution. AIAA J. **6**(7), 1378–1379 (1968)

Chapter 3
Transfer Between Libration Point Orbits and Lunar Orbits in Earth-Moon System

This chapter is devoted to the study of the transfer problem from a libration point orbit of the Earth-Moon system to an orbit around the Moon. The transfer procedure analysed has two legs: the first one is an orbit of the unstable manifold of the libration orbit and the second one is a transfer orbit between a certain point on the manifold and the final lunar orbit. There are only two manoeuvres involved in the method and they are applied at the beginning and at the end of the second leg. Although the numerical results given in this paper correspond to transfers between halo orbits around the L_1 point (of several amplitudes) and lunar polar orbits with altitudes varying between 100 and 500 km, the procedure we develop can be applied to any kind of lunar orbits, libration orbits around the L_1 or L_2 points of the Earth-Moon system, or to other similar cases with different values of the mass ratio.

3.1 Introduction

Since some years ago, missions from different space agencies have revisited the Moon with several kinds of spacecrafts and devices including lunar orbiters, landers, rovers, or sample return spacecraft. A lunar L_1 or L_2 Gateway Station can support infrastructures beyond orbits around the Earth and serve as a staging location for missions to the Moon (see Ref. [14]).

Several approaches have been used for the analysis of these missions, depending on the different goals to be achieved (see, for instance, [3, 5, 8, 16, 17, 19]). These include the acquisition of accurate and high-resolution 3D maps of the Moon's surface for the selection of future landing sites, the exploration of possible water resources near its poles, testing new technologies, etc. Some of these lunar missions, such as Artemis or Chang'e 2, have used libration point orbits (LPO) as their nominal trajectory or have visited this kind of orbits in the Earth-Moon and the Sun-Earth systems. As is well known, in the circular restricted 3-body problem (CR3BP), or

J. Yuan et al., *Low Energy Flight: Orbital Dynamics and Mission Trajectory Design*, https://doi.org/10.1007/978-981-13-6130-2_3

in any perturbed form of it, libration point orbits around the L_1 and L_2 points can be easily reached from the vicinity of the small primary. This fact is because the stable manifolds of these libration point orbits have close approaches to the small primary or even intersect it. Examples of this applications are the Earth departure to the LPOs in the Sun-Earth system, or the Moon departure to the LPOs in the Earth-Moon system.

Furthermore, since there exist heteroclinic connections between the LPOs of the Sun-Earth and the Earth-Moon systems, it is possible to transfer from one system to the other [12, 13], or in other scenarios, to obtain low energy transfer paths to visit other Solar System bodies. This kind of connections was used by the Chinese Chang'e 2 spacecraft [6, 15]. Chang'e 2 was launched in October 2010 to conduct research from a 100 km high lunar polar orbit and provided high-resolution images of the lunar surface. In June 2011 Chang'e 2 left from the lunar orbit towards the Sun-Earth L_2 Lagrangian point and reached it after a 77-day cruise. Although it was originally expected to remain at L_2 until the end of 2012, in April 2012 Chang'e 2 started an extended mission to flyby the asteroid 4179 Toutatis at a distance of 3.2 km in December of the same year [6].

In the framework of CR3BP, we analyse the transfer from a LPO around a collinear equilibrium point to an orbit around the small primary. The analysis of these trajectories, which in some sense are the inverse of the transfers usually considered in the literature [18], enhance the possibilities to be considered for missions relating LPOs.

The procedure is developed and explained in a general context, but most of the computations are done for the transfer case between halo orbits with different z-amplitudes around the L_1 point of the Earth-Moon system, and polar orbits around the Moon (taking into account the orientation of the rotation axis of the Moon) with altitudes varying between 100 and 500 km. Nevertheless, it can also be applied to compute transfers from any kind of LPOs, around L_1 or L_2, of an arbitrary CR3BP model, to circular Keplerian orbits around the small body of the system.

In order to optimize the transfer strategy, it is convenient to depart at a point along the unstable manifold of the LPO at a distance from the Moon between 45,000 and 20,000 km and to perform the orbit inclination change manoeuvre to reach the desired lunar orbit inclination at this point. The size of the departing halo orbit has little influence on the total cost, and of course the cost depends on and decreases when increasing the altitude of the target lunar orbit. In order to shorten the time of flight, the departure manoeuvre should be done at the first apolune of the unstable invariant manifold of the LPO.

In this chapter we do not address the cost of the transfer in a high realistic force model (including, for instance, the gravitational effect of the Sun or the eccentricity of the Moon's orbit) or the influence of the departing epoch from the LPO. We have checked that modulus of the relative residual accelerations between the CR3BP and a n-body model (defined by the JPL ephemerides), which is of the order of 0.0012, independently of the transfer orbit and the initial epoch. So the qualitative and quantitative results for real transfers should be close to the ones given in this chapter, with estimated variations in the total cost of less than 10%.

The chapter is organized as follows: in Sect. 3.2 the reference model, changes of coordinates and the notation used are introduced. In Sect. 3.3 we explain the transfer procedure and how the two transfer manoeuvres are computed. The numerical results for the transfer between halo orbits of the Earth-Moon system and lunar polar orbits are given in Sect. 3.4. Finally, the last Section concludes the study.

3.2 The Dynamic Model

3.2.1 Equations of Motion

Along this paper, we use the CR3BP as the dynamic model for the motion of a spacecraft in the Earth-Moon system. That is: with the mass ratio μ equal to 0.012150582 (according to the DE401 ephemeris data file). In the synodic reference frame, the origin is set at the barycentre of the system, the positive x-axis is pointing from the Moon to the Earth, the z-axis is in the direction of the normal vector to the Moon's orbit around the Earth, and the y-axis completes the right-hand coordinate system. In the usual non-dimensional system, in which the length unit is the semi-axis of the Moon's orbit around the Earth and the time unit is such that the period of the Moon around the Earth is 2π, one unit of non-dimensional distance is 0.3844037×10^6 km and one time unit is approximately 0.377496×10^6 s.

According to Ref. [22], the differential equations of the model can be written as

$$\begin{aligned} \ddot{x} - 2\dot{y} &= \Omega_x, \\ \ddot{y} + 2\dot{x} &= \Omega_y, \\ \ddot{z} &= \Omega_z, \end{aligned} \tag{3.1}$$

where $\Omega(x, y, z) = (x^2 + y^2)/2 + (1 - \mu)/r_1 + \mu/r_2 + \mu(1 - \mu)/2$, and

$$r_1 = \sqrt{(x - \mu)^2 + y^2 + z^2}, \quad r_2 = \sqrt{(x - \mu + 1)^2 + y^2 + z^2}.$$

The CR3BP has a first integral given by $\mathscr{C} = 2\Omega(x, y) - \dot{x}^2 - \dot{y}^2 - \dot{z}^2$, where $\mathscr{C}$ is the so called Jacobi constant.

It is well known that the CR3BP differential equations (3.1) have three collinear equilibrium (or libration) points, L_1, L_2 and L_3, and the two equilateral ones, L_4 and L_5. If x_{L_i} denotes the x coordinate of the L_i point, then $x_{L_2} \le \mu - 1 \le x_{L_1} \le \mu \le x_{L_3}$.

According to the values of the eigenvalues associated with the collinear equilibrium points, $\pm i\lambda_1$, $\pm i\lambda_2$, $\pm\lambda_3$, with $\lambda_{1,2,3} \in \mathbb{R}$ (see Ref. [22]), the three points are centre $\times$ centre $\times$ saddle critical points. Due to the centre $\times$ centre part (associated with the imaginary eigenvalues $\pm i\lambda_1$ and $\pm i\lambda_2$), and considering all the energy levels, there are 4-dimensional centre manifolds around these points that, for a given value of the Jacobi constant, are just 3D sets. The saddle component of the flow (associated with the real eigenvalues $\pm\lambda_3$) constitutes the stable and unstable invariant manifolds of the unstable point. For values of the Jacobi constant close to that of L_1 and L_2, the unstable and stable manifolds associated with the periodic orbits and invariant tori restrict the dynamics in the neighbourhood of the equilibrium points (see Refs. [11, 14]).

3.2.2 *Change of Coordinates Between the Synodic CR3BP and the Moon-Centred Sidereal Frames*

In this section we briefly describe the change of coordinates that has been used to transform from the usual synodic coordinates of the CR3BP to a Moon-centred sidereal frame $\{M - x'y'z'\}$ in which the Keplerian orbital elements are $(a, i, e, \Omega, \omega, f)$ are described. For this purpose, we introduce an intermediate sidereal reference frame $\{M - \overline{x}\,\overline{y}\,\overline{z}\}$, which is aligned with $\{O - xyz\}$ at an initial epoch t_0, that is set as $t_0 = 0$.

Let $(x^{syn}, y^{syn}, z^{syn})$ be the synodic coordinates of a point $\boldsymbol{r}^{syn}$ in the CR3BP synodic reference frame, $\{O - xyz\}$, with origin O at the Earth-Moon barycentre (see Fig. 3.1). A translation $\boldsymbol{d}$ along the x-axis of modulus $1 - \mu$ sets the origin at the Moon, and a rotation around the z axis of angle $nt = t$, which is expressed as $R_3(t)$, transforms the synodic coordinates to the intermediate sidereal coordinates in the reference frame $\{M - \overline{x}\,\overline{y}\,\overline{z}\}$. In this way, the transformation between the synodic and the sidereal system is given by:

$$\boldsymbol{r}^0 = R_3(t)(\boldsymbol{r}^{syn} + \boldsymbol{d}) = \begin{pmatrix} \cos(t) & -\sin(t) & 0 \\ \sin(t) & \cos(t) & 0 \\ 0 & 0 & 1 \end{pmatrix} \begin{pmatrix} x^{syn} + 1 - \mu \\ y^{syn} \\ z^{syn} \end{pmatrix}.$$

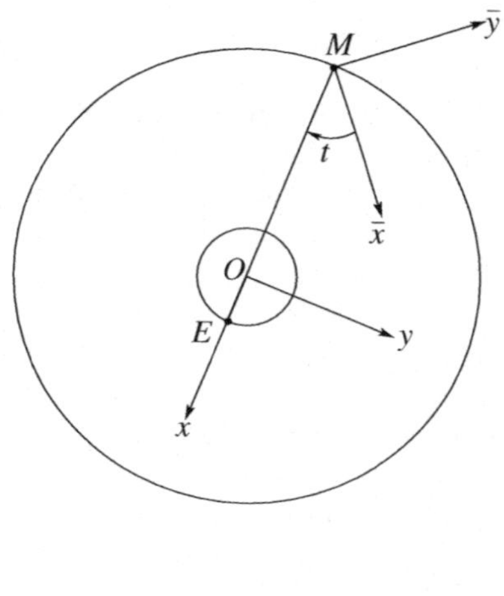

Fig. 3.1 Relation between the usual CR3BP synodic reference frame $\{O - xyz\}$, and the intermediate Moon-centred sidereal reference frame $\{M - \overline{x}\,\overline{y}\,\overline{z}\}$. The z axis of both systems is perpendicular to the plane displayed

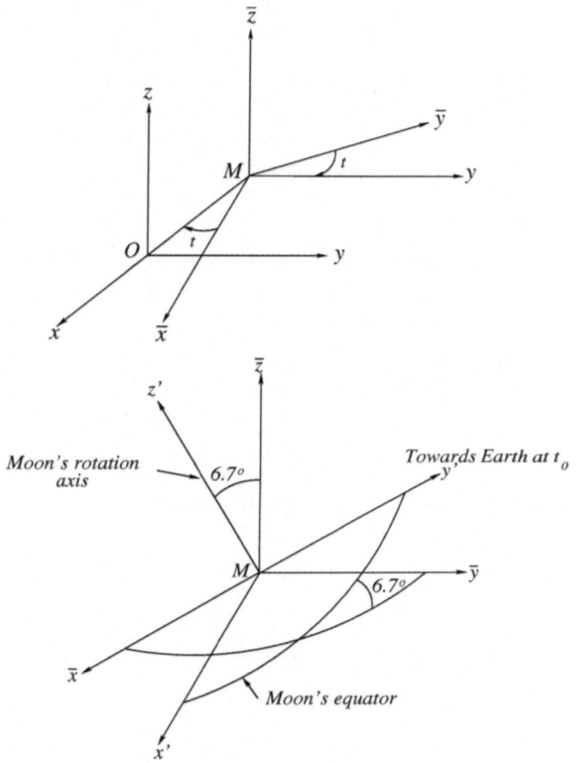

Fig. 3.2 Moon-centred sidereal reference frames. The top plot displays the relation between the $\{M - \bar{x}\bar{y}\bar{z}\}$ reference system and the usual CR3BP synodic reference frame $\{O - xyz\}$. Both frames are parallel at $t_0 = 0$. The bottom plot displays the two inertial Moon-centred reference frames: $\{M - \bar{x}\bar{y}\bar{z}\}$ and $\{M - x'y'z'\}$. The inclination i of an orbit around the Moon is measured w.r.t. the (x', y') plane

Usually, the inclination i of the orbit of a spacecraft around the Moon is measured w.r.t. the Moon's equator. For this purpose, we introduce the Mean Earth/Polar Axis Lunar reference system $\{M - x'y'z'\}$, which, according to Ref. [1] (see Fig. 3.2), is defined as follows: the origin is located at the centre of the Moon, the z'-axis is aligned with the mean Moon's rotation axis (orthogonal to the Moon's equator) at a certain epoch t_0, and the Prime Meridian (0° lunar longitude) is the mean Earth direction at the same epoch. So the x'-axis points towards the mean Earth at t_0, and the y'-axis completes the right-hand coordinate system. Due to the inclination of the Moon's equator with respect to the Moon's orbital plane around the Earth ($(\bar{x}, \bar{y})$ plane), the z' axis forms an angle of approximate 6.7° with the $\bar{z}$-axis. Due to the long precession period of the Moon's rotation axis (of about 18.6 years) we assume that the z'-axis is fixed during the mission lifetime.

To compute the change of coordinates between the intermediate sidereal reference frame $\{M - \bar{x}\bar{y}\bar{z}\}$ and the sidereal reference frame $\{M - x'y'z'\}$, we use an additional frame, *ICRF*, which is the one used in the JPL ephemeris file DE421. If $\boldsymbol{r}^{sid}$ and $\boldsymbol{r}^{icrf}$ denote the coordinates of a point in the reference frames $\{M - x'y'z'\}$ and *ICRF*, respectively, the relation between them is

$$r^{sid} = B\,r^{icrf},$$

where the matrix B depends on the the lunar libration angles at the initial epoch t_0 (see Ref. [4] for details).

If r^0 denotes the coordinates of the point r^{icrf} in the $\{M - \overline{x}\,\overline{y}\,\overline{z}\}$ frame, then, according to Ref. [4], we have

$$r^{icrf} = C\,r^0,$$

where the columns of the matrix C are: r_m, $(r_m \times v_m) \times v_m$, and $r_m \times v_m$. The vectors r_m and v_m are the position and velocity of the Moon relative to the Earth at the epoch t_0, respectively (and can be also obtained from DE421).

In this way, we get the transformation between the inertial reference frames $\{M - x'y'z'\}$ and the synodic reference frame $\{O - x\,y\,z\}$ as

$$r^{sid} = B\,C\,r^0 = B\,C\,R_3(t)(r^{syn} + d) = T(r^{syn}),$$

where the transformation T is defined as

$$T(r^{syn}) := A \cdot (r^{syn} + d).$$

The matrix $A = B\,C\,R_3(t)$ depends on the initial epoch. Using the data of the ephemeris file DE421, and as initial reference date t_0 the epoch 2020 Jan. 1 12:00:00(TDB)

$$A = \begin{pmatrix} 0.99338553 & -0.00636032 & -0.11465040 \\ 0.00250881 & 0.99942862 & -0.03370651 \\ 0.11479928 & 0.03319592 & 0.99283390 \end{pmatrix}. \tag{3.2}$$

These are the values that have been used for the numerical results given in Sect. 3.4. Although we have not done a systematic study of the results as a function of the initial epoch, i.e., the transformation A, some explorations suggest that the variations in the Δv costs are small (usually less than 7%).

To simplify the notation, from now on we will avoid the use of the super-index *sid* and *syn* when the context is clear enough.

3.3 Computation of a Transfer from LPO to a Circular Lunar Orbit

In this section we explain the general procedure used to compute transfer trajectories from a LPO around the L_1 point of the Earth-Moon system (such as a halo orbit) to a circular lunar orbit with a certain inclination i. For the explanation of the method

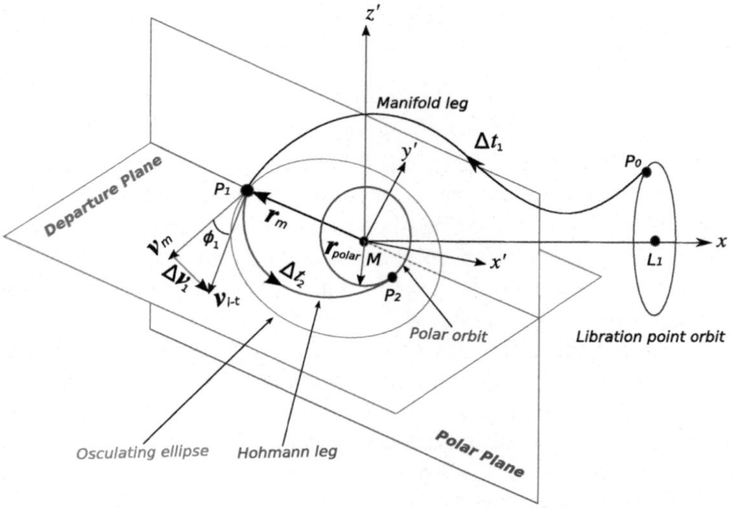

Fig. 3.3 Sketch of whole transfer process. The transfer starts at the point P_0 of a LPO. From P_0 to P_1 (manifold leg) it follows an orbit of the unstable manifold of the LPO. The first transfer manoeuvre $\Delta \boldsymbol{v}_1$ is performed at P_1 at a distance r_m from the Moon. At this point the spacecraft is injected in the Hohmann leg contained in the plane of the final target orbit. The second transfer manoeuvre, not represented in the figure, is done at P_2, where the spacecraft is injected in a lunar polar orbit of radius r_{polar}

we assume that the LPO and the orbit around the Moon (determined by its radius) are fixed.

In the procedure, the transfer trajectory has two legs. The first one, *the manifold leg*, goes from P_0 to P_1 and follows an orbit of the unstable manifold of the LPO (in fact, of the branch of the manifold that goes towards the Moon). The second one, *the Hohmann leg*, goes from P_1 to P_2, and is a Hohmann like transfer orbit that, after a suitable manoeuvre $\Delta \boldsymbol{v}_1$ at P_1 connects a point of the unstable manifold with the lunar orbit at P_2, where it does an insertion manoeuvre $\Delta \boldsymbol{v}_2$. P_1 and P_2 are the apocentre and pericentre of the Hohmann leg, respectively. The whole transfer process is sketched in Fig. 3.3.

Assuming that no Δv is required to depart from the LPO, due to the fact that the inherent instability of the orbit does this task for us, the full process requires two manoeuvres $\Delta \boldsymbol{v}_1$ and $\Delta \boldsymbol{v}_2$. The first one is required to depart from the orbit of the unstable manifold, and we call it the *departure manoeuvre*. The second manoeuvre $\Delta \boldsymbol{v}_2$, namely the *insertion manoeuvre*, inserts the spacecraft into an orbit around the Moon. In principle, the final goal is to minimise the total transfer cost in terms of $\Delta v_{tot} = \|\Delta \boldsymbol{v}_1\| + \|\Delta \boldsymbol{v}_2\|$. In this chapter we explore the role of the free parameters of the problem in the value of Δv_{tot}.

We assume that we move along an orbit of the unstable manifold of the departure LPO during Δt_1 time units, reaching the state $P_1 = (\boldsymbol{r}_m^{syn}, \boldsymbol{v}_m^{syn})$ in the synodic system

$\{O - xyz\}$. At this point, after transforming it to the sidereal system $\{M - x'y'z'\}$, we perform the first manoeuvre $\Delta \boldsymbol{v}_1$ to reach the polar lunar orbit by means of a non-coplanar Hohmann transfer.

To compute $\Delta \boldsymbol{v}_1$, we need to look at P_1 in a two-body problem (2BP) framework around the Moon. Let $(\boldsymbol{r}_m, \boldsymbol{v}_m)$ be the coordinates of this point in sidereal reference frame $\{M - x'y'z'\}$, computed using the transformation T between the sidereal and the synodic systems, together with its derivative. Note that, the position vector $\boldsymbol{r}_m$ is aligned with the intersection line between the departure plane and the polar plane.

The point $(\boldsymbol{r}_m, \boldsymbol{v}_m)$ defines a Keplerian osculating orbit around the Moon (represented by the purple line in Fig. 3.3 contained in the departure plane) whose associated Keplerian elements are $(a_m, i_m, e_m, \Omega_m, w_m, f_m)$. In general, the inclination i_m is not the one of the target lunar orbit. The computation of the suitable $\Delta \boldsymbol{v}_1$ which performs the transfer to a lunar orbit with a given inclination i is explained later.

After the first manoeuvre, $\Delta \boldsymbol{v}_1$, the state vector becomes $(\boldsymbol{r}_{i-t} = \boldsymbol{r}_m, \boldsymbol{v}_{i-t} = \boldsymbol{v}_m + \Delta \boldsymbol{v}_1)$, which is the initial state of the transfer orbit (represented by blue line in Fig. 3.3 contained in the polar plane). This point is propagated during Δt_2 time units until it reaches a sphere centred at the Moon with radius r_{polar}. In the $\{O - xyz\}$ reference system the sphere is defined as

$$S = \left\{ (x, y, z) \mid (x + 1 - \mu)^2 + y^2 + z^2 - r_{polar}^2 = 0 \right\}.$$

At this point the second manoeuvre $\Delta \boldsymbol{v}_2$ is performed in order to get captured into a circular orbit around the Moon with the desired inclination. All the propagations are done in the synodic reference frame using the CR3BP Eq. (3.1).

In conclusion, the problem is essentially a non-coplanar Hohmann transfer where the locations of the two manoeuvres are not specified in advance. In a first approximation we assume that these locations are the apocentre and pericentre of a non-coplanar Hohmann-like transfer ellipse, from a certain point of the unstable manifold to a non-coplanar circular orbit around the Moon of radius r_{polar}. As it has already been said, the main objective is to determine where and how to perform the two manoeuvres in order to minimise $\Delta v_{tot} = \|\Delta \boldsymbol{v}_1\| + \|\Delta \boldsymbol{v}_2\|$.

3.3.1 *Computation of the Transfer Manoeuvre to a Keplerian Ellipse with a Fixed Inclination i*

As it has been stated, the first manoeuvre deals with the change of inclination between the osculating ellipse associated with a certain point P_1 of the unstable manifold and the circular lunar orbit. In this section we explain the general method to compute the manoeuvre that injects the spacecraft into a Keplerian ellipse contained in a plane with a fixed inclination i. The method is similar to the one developed in Refs. [7, 20].

In general, to guarantee that a spacecraft with initial conditions $(\boldsymbol{r}, \boldsymbol{v})$ moves on a given target plane defined by its normal vector $\boldsymbol{n}$, its initial velocity $\boldsymbol{v}$ must be

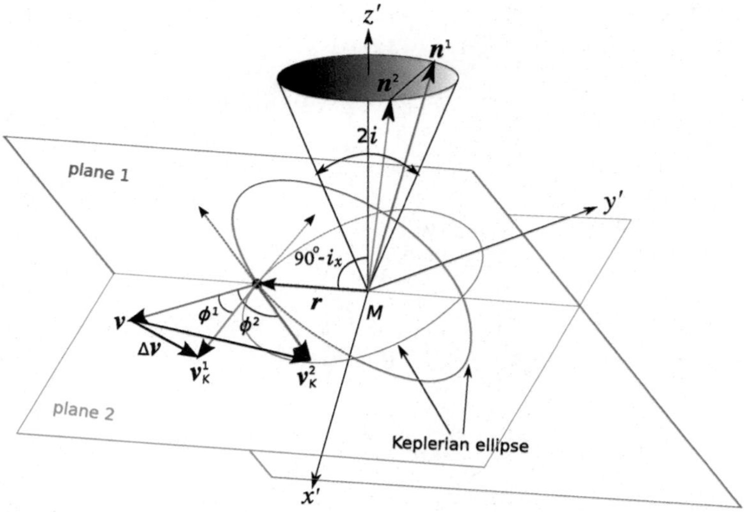

Fig. 3.4 Intersection of the cone of amplitude $2i$ around the z' axis and the plane perpendicular to $\boldsymbol{r}$. The intersecting directions $\boldsymbol{n}^1$ and $\boldsymbol{n}^2$ define the two planes shown in the figure. We display here the case when $i \in (i_x, 90°)$ for which we have two possible $\boldsymbol{v}_K^{1,2}$. For $i \in (90, 180° - i_x)$ the velocities are the two dashed arrows aligned with $\boldsymbol{v}_K^{1,2}$. Note that when $i = i_x$ (or $i = 180° - i_x$), $\boldsymbol{n}^1$ and $\boldsymbol{n}^2$, as well as the two velocities $\boldsymbol{v}_K^{1,2}$, coincide, and when $i = 90°$, $\boldsymbol{v}_K^{1,2}$ are aligned but pointing in opposite directions

contained in the plane, this is $\boldsymbol{v} \perp \boldsymbol{n}$. Next we are going to explain how to compute the modulus of $\boldsymbol{v}$ and its direction.

The inclination i of the target plane is the angle between the normal to this plane $\boldsymbol{n}$ and the z'-axis. Therefore, $\boldsymbol{n}$ is on a cone of angle $2i$ around z' axis with vertex at the Moon (see Fig. 3.4). Moreover, $\boldsymbol{n}$ is also in a plane perpendicular to $\boldsymbol{r}$. As a consequence, it is the intersection of both surfaces, which in general, are two lines defined by the unitary vectors $\boldsymbol{n}^1$ and $\boldsymbol{n}^2$.

Let $\boldsymbol{n} = (n_x, n_y, n_z)$ be a unitary vector perpendicular to $\boldsymbol{r} = (x, y, z)$, so $\boldsymbol{r} \cdot \boldsymbol{n} = 0$. We require the velocity $\boldsymbol{v}_K$ of the Keplerian orbit departing from $\boldsymbol{r}$ to be in the plane defined by $\boldsymbol{n}$. The modulus of $\boldsymbol{v}_K$ can be determined by the vis-viva equation for a Keplerian ellipse with semi-major axis a,

$$v_K = \sqrt{2\frac{GM}{r} - \frac{GM}{a}}, \tag{3.3}$$

where G is the gravitational constant, M is the mass of the Moon and $r = \|\boldsymbol{r}\|$.

If $\boldsymbol{r}$ is the apocentre or pericentre of the ellipse, $\boldsymbol{v}_K$ must be perpendicular to $\boldsymbol{n}$ and $\boldsymbol{r}$, so

$$\boldsymbol{v}_K = v_K \frac{\boldsymbol{n} \times \boldsymbol{r}}{\|\boldsymbol{n} \times \boldsymbol{r}\|}. \tag{3.4}$$

In summary, the conditions that must be fulfilled by a unitary vector $\boldsymbol{n}$ orthogonal to the plane containing $\boldsymbol{r}$ and $\boldsymbol{v}_K$ are:

$$\boldsymbol{r}^T \cdot \boldsymbol{n} = 0, \quad \boldsymbol{n} \cdot z' = \cos i, \quad \boldsymbol{n}^T \cdot \boldsymbol{n} = 1,$$

from which we get:

$$n_x = \frac{-b \pm \sqrt{b^2 - 4ac}}{2a}, \quad n_y = -\frac{n_x x + z\cos i}{y}, \quad n_z = \cos i,$$

where, $a = x^2 + y^2$, $b = 2xz\cos i$, $c = z^2\cos^2 i - y^2\sin^2 i$. Clearly, we must require that $b^2 - 4ac \geq 0$, from which it follows that:

$$\sin^2 i \geq \frac{z^2}{x^2 + y^2 + z^2} = \sin^2 i_x \;\Rightarrow\; \cos(2i) \leq \cos(2i_x) \;\Rightarrow\; i_x \leq i \leq 180^\circ - i_x, \tag{3.5}$$

where i_x in the above inequality (3.5) is the complementary of the angle between the position vector $\boldsymbol{r}$ and the z'-axis, which is also the angle between $\boldsymbol{r}$ and the (x', y') plane.

When the inclinations i and i_x do not fulfil the above inequalities, there are no intersections between the cone and the plane perpendicular to $\boldsymbol{r}$ (see Fig. 3.4 for $i \in [i_x, 90^\circ]$). Therefore, the available range of values for i is $[i_x,\ 180^\circ - i_x]$.

In general, for a fixed value of $i \in [i_x, 180^\circ - i_x]$, there are two solutions for $\boldsymbol{n}$ and, according to Eq. (3.4), we have two different $\boldsymbol{v}_K$, in the plane defined by $\boldsymbol{n}^1$ (plane 1 in Fig. 3.4) and by $\boldsymbol{n}^2$ (plane 2 in Fig. 3.4), respectively. The required manoeuvres for the transfer are given by:

$$\Delta\boldsymbol{v} = \boldsymbol{v}_K - \boldsymbol{v}. \tag{3.6}$$

We note that, if $i = i_x$ (or $180^\circ - i_x$), the two velocities $\boldsymbol{v}_K^1$ and $\boldsymbol{v}_K^2$ coincide and thus we only have one possible $\Delta\boldsymbol{v}$. Also when $i = 90^\circ$, $\boldsymbol{v}_K^1$ and $\boldsymbol{v}_K^2$ are aligned but point in opposite directions.

3.3.2 *Computation of the Departure Manoeuvre: First Approximation*

Next we are going to apply the method described in the previous section to determine a first approximation of the first manoeuvre $\Delta\boldsymbol{v}_1$.

The first manoeuvre is performed at the apocentre of the (Hohmann) transfer ellipse, so the modulus of the velocity after the manoeuvre must be

$$v_{i-t} = \sqrt{2\frac{GM}{r_m} - 2\frac{GM}{r_{polar} + r_m}}, \tag{3.7}$$

where G is the gravitational constant, M is the mass of the Moon. We have used that the pericentre of the ellipse is on the polar orbit around the Moon with radius r_{polar}, so the semi-major axis of this ellipse is $a_m = (r_{polar} + r_m)/2$.

Since we want to reach a polar orbit around the Moon, the final inclination must equal 90°, however, the orbit inclination of the transfer ellipse at the point P_1 (denoted as i_1 here) does not necessarily have to equal 90°. So we have to explore all the suitable values $i_1 \in [i_x,\ 180° - i_x]$ of the inclination of the transfer ellipse. We have seen that for each value of the orbit inclination, there are two possible values of $\boldsymbol{n}$ and, as a consequence, two feasible velocities $\boldsymbol{v}_{i-t}^{1,2}$, both of which must be considered for the determination of the first manoeuvre $\Delta\boldsymbol{v}_1 = \boldsymbol{v}_{i-t}^{1,2} - \boldsymbol{v}_m$. Note that the computation of $\Delta\boldsymbol{v}_1$ is done in the sidereal system $\{M - x', y', z'\}$, and also that its modulus is the same in both the sidereal and the synodic system. This is because the transformation between both systems is orthogonal. The two values $\boldsymbol{v}_{i-t}^{1,2}$, that will be refined later, are used as the initial seed for the computation of the departure manoeuvre $\Delta\boldsymbol{v}_1$.

Denote ϕ_1 as the angle between the two sidereal velocities at P_1, $\boldsymbol{v}_m$ and $\boldsymbol{v}_{i-t}$ (displayed in Fig. 3.3). In a first approximation, the modulus of the departure manoeuvre Δv_1 can be written as:

$$\Delta v_1 = \sqrt{v_m^2 + v_{i-t}^2 - 2v_m v_{i-t} \cos\phi_1}, \tag{3.8}$$

where $v_m = \sqrt{2\frac{GM}{r_m} - \frac{GM}{a_m}}$, and P_1 is considered as a point in the osculating ellipse with semi-major axis a_m, as defined before. From Eq. 3.8 it follows that the main factors to be considered for the determination of v_m, v_{i-t} and ϕ_1 are:

1. The point P_1
   - The distance from P_1 to the centre of the Moon r_m. According to Eq. (3.7), for a fixed polar radius, v_{i-t} depends only on this distance.
   - The inclination of the osculating ellipse at P_1, which in fact affects the change of inclination $\Delta i_1 = |i_1 - i_m|$, and thus ϕ_1.
   - The velocity $\boldsymbol{v}_m$ of the spacecraft at P_1.
2. The inclination of the transfer ellipse i_1, which essentially affects ϕ_1.
3. The departing LPO, since P_1 is in the unstable manifold associated with it.
4. The orbit selected for the manifold leg on the unstable manifold.

We recall that the modulus of $\boldsymbol{v}_{i-t}$ is specified by the vis-viva Eq. (3.7), but its direction varies with i_1 and, subsequently, the angle ϕ_1 varies. Of course, ϕ_1 depends also on the inclination of the osculating ellipse at P_1. It must be noted that ϕ_1, although it is related to the angle between the departure and the polar planes, is not exactly equal to this angle. This is due to the fact that the manifold leg is not a 2BP orbit, so the osculating inclination along it varies and, depending on the point P_1 the difference between ϕ_1 and Δi_1 can be large.

3.3.3 Refinement of the Departure Manoeuvre and Determination of P_2

As mentioned before, the insertion manoeuvre at the point P_2 is done at a point on the sphere S previously defined, and it is also at the pericentre of the transfer ellipse (in a two-body problem scenario). We denote by $\boldsymbol{r}_{f-t}$ and $\boldsymbol{v}_{f-t}$ the sidereal position and velocity of the Hohmann ellipse at its pericentre. By means of the inverse of the transformation between the sidereal and synodic systems, together with its derivative, we obtain the corresponding position and velocity in the synodic system: $\boldsymbol{r}_{f-t}^{syn}$ and $\boldsymbol{v}_{f-t}^{syn}$.

To fulfil the above two conditions, the following equations must be satisfied:

$$\boldsymbol{v}_{f-t}^{syn} \cdot \nabla S = 0, \quad \boldsymbol{r}_{f-t}^{syn} \cdot \boldsymbol{v}_{f-t}^{syn} = 0,$$

where $S(x, y, z) = (x + 1 - \mu)^2 + y^2 + z^2 - r_{polar}^2$. We rewrite the second equation as $g = \boldsymbol{r}_{f-t}^{syn} \cdot \boldsymbol{v}_{f-t}^{syn}$.

The refinement of the first manoeuvre is iterative, it has two steps and provides also the point P_2 where the second manoeuvre is done. The first step is to integrate the initial seed given by the Hohmann transfer once it has been converted into the synodic system, $P_1 = (\boldsymbol{r}_{i-t}^{syn}, \boldsymbol{v}_{i-t}^{syn})$, until the condition $g = \boldsymbol{r}^{syn} \cdot \boldsymbol{v}^{syn} = 0$ is fulfilled. This gives us Δt_2, the approximate time of flight along the transfer orbit and a final point $P_2 = (\boldsymbol{r}_{f-t}^{syn}, \boldsymbol{v}_{f-t}^{syn})$. To have $g = 0$ (with an accuracy of 10^{-12}) we integrate until we get the first change of sign of g, and then we apply Newton's method by adjusting the value of the transfer time Δt_2.

Since we are integrating the initial condition $(\boldsymbol{r}_{i-t}^{syn}, \boldsymbol{v}_{i-t}^{syn})$ using the CR3BP equations, in general, when we arrive at P_2 after Δt_2 time units

$$S(\boldsymbol{r}_{f-t}^{syn}) \neq 0,$$

because $\boldsymbol{r}_{f-t}^{syn}$ has been computed in a two-body problem approximation. Note that, in fact, $\boldsymbol{r}_{f-t}^{syn} = \boldsymbol{r}_{f-t}^{syn}(\boldsymbol{r}_{i-t}^{sid}, \boldsymbol{v}_{i-t}^{sid})$, so

$$S(\boldsymbol{r}_{f-t}^{syn}) = S\left(\boldsymbol{r}_{f-t}^{syn}(\boldsymbol{r}_{i-t}^{sid}, \boldsymbol{v}_{i-t}^{sid})\right) = \hat{S}(\boldsymbol{r}_{i-t}^{sid}, \boldsymbol{v}_{i-t}^{sid}),$$

and the above condition can be also written as

$$\hat{S}(\boldsymbol{r}_{i-t}^{sid}, \boldsymbol{v}_{i-t}^{sid}) \neq 0.$$

Now we want to modify the initial condition at P_1 keeping the position $\boldsymbol{r}_{i-t}^{sid}$ fixed as well as the direction of $\boldsymbol{v}_{i-t}^{sid}$ in order that the transfer ellipse remains in a transfer plane with the desired inclination. Then, if we remove those arguments of $\hat{S}$ that are kept fixed, in fact we need to solve

$$\hat{S}(\boldsymbol{v}_{i-t}^{sid}(1+\Delta v)) = 0,$$

where now the only free parameter is Δv.

The value of Δv can be obtained using Newton's method, from which we get

$$\Delta v = -\frac{\hat{S}(\boldsymbol{v}_{i-t}^{sid})}{D\hat{S}_{v_x} v_x + D\hat{S}_{v_y} v_y + D\hat{S}_{v_z} v_z},$$

where $\boldsymbol{v}_{i-t}^{syn} = (v_x, v_y, v_z)$ and $D\hat{S}_{v_x}$, $D\hat{S}_{v_y}$ and $D\hat{S}_{v_z}$ are components of the matrix $D\hat{S}$ (see Ref. [2]). Using $\boldsymbol{X}_i^{syn} = (\boldsymbol{r}_{i-t}^{syn}, \boldsymbol{v}_{i-t}^{syn})$, $\boldsymbol{X}_f^{syn} = (\boldsymbol{r}_{f-t}^{syn}, \boldsymbol{v}_{f-t}^{syn})$ one can write

$$D\hat{S} = \left[\frac{\partial S}{\partial \boldsymbol{r}_f^{syn}} \cdot (\Phi + F \cdot Dt) \cdot \frac{\partial \boldsymbol{X}_i^{syn}}{\partial \boldsymbol{X}_i^{sid}} \right],$$

where $\Phi = \dfrac{\partial \boldsymbol{X}_f^{syn}}{\partial \boldsymbol{X}_i^{syn}}$ is given by the state transition matrix of CR3BP, $F = \dfrac{\partial \boldsymbol{X}_f^{syn}}{\partial t_f}$ is the CR3BP vector-field, $Dt = \dfrac{\partial \Delta t_2}{\partial \boldsymbol{X}_i^{syn}} = -\dfrac{Dg \cdot \Phi}{Dg \cdot F}$, and

$$\frac{\partial \boldsymbol{X}_i^{syn}}{\partial \boldsymbol{X}_i^{sid}} = \left(\begin{array}{c|c} A & 0 \\ \hline A \cdot J & A \end{array} \right),$$

where the matrix A is given by Eq. (3.2), and $J = \dot{R}_3(\Delta t_1)$.

Once the value of Δv has been computed, after one iteration of Newton's method, we go to the first step. The iterative procedure finishes when the value of $\hat{S}$ is less than a certain tolerance (that we have fixed equal to 10^{-12}).

3.3.4 Computation of the Insertion Manoeuvre at P_2

The aim of the insertion manoeuvre $\Delta \boldsymbol{v}_2$ is to move from the transfer ellipse previously determined to a circular polar orbit of radius r_{polar} around the Moon. It must be noted that at P_2 the vector $\boldsymbol{r}_{f-t}^{syn} \times \boldsymbol{v}_{f-t}^{syn}$ is not, in general, orthogonal to the polar plane, so $\Delta \boldsymbol{v}_2$ accounts for this (small) correction of inclination together with the insertion into a polar orbit. The general procedure to do this manoeuvre is the one that has been explained in Sect. 3.3.1. We follow it to compute $\Delta \boldsymbol{v}_2$, recalling that we obtain two solutions for $\boldsymbol{v}_c^{sid}$ with the same modulus and opposite directions. Then, we choose the one for which Δv_2 is minimum. It must also be noted that the location of P_2 can be constrained by mission requirements or by a suitable value of the argument of the ascending node Ω of the orbit; in this paper almost all the possibilities have been explored in order that the most suitable selection, according to the constraints, can be done.

The modulus of the velocity in a circular polar orbit is given by:

$$v_c^{sid} = \sqrt{\frac{GM}{r_{polar}}}, \tag{3.9}$$

and its inclination with respect to the Moon's equator is $i_2 = 90°$.

The value of Δv_2 is obtained from

$$\Delta \boldsymbol{v}_2 = \boldsymbol{v}_c^{sid} - \boldsymbol{v}_{f-t}^{sid}. \tag{3.10}$$

and, as in the previous case, the procedure can be summarised as:

$$\boldsymbol{v}_{f-t}^{syn} \xrightarrow{T} \boldsymbol{v}_{f-t}^{sid} \xrightarrow{\Delta \boldsymbol{v}_2} \boldsymbol{v}_c^{sid}$$

Similar to what has been said about the factors that influence the value of the manoeuvre Δv_1 at P_1, we denote by ϕ_2 the angle between the two sidereal velocities at P_2, $\boldsymbol{v}_{f-t}$ and $\boldsymbol{v}_c$. Then, the modulus of the insertion manoeuvre Δv_2 can be written as:

$$\Delta v_2 = \sqrt{v_c^2 + v_{f-t}^2 - 2 v_c v_{f-t} \cos \phi_2}, \tag{3.11}$$

where, as explained in Sect. 3.3.1, P_2 is the pericentre of the transfer ellipse, which gives us an approximate value

$$v_{f-t} = \sqrt{2\frac{GM}{r_{polar}} - \frac{GM}{r_{polar} + r_m}}.$$

Although it is refined later by the procedure, it provides a rough estimate of Δv_2 that mainly depends on r_{polar}, r_m and ϕ_2.

If $\boldsymbol{r}_{f-t}^{syn} = (x, y, z)$ denotes the coordinates of the position of P_2 in the synodic reference frame $\{M - xyz\}$, then its longitude β and latitude λ w.r.t. the Moon, are given by

$$\beta = \arctan \frac{y}{x + 1 - \mu}, \quad \lambda = \arctan \frac{z}{\sqrt{(x + 1 - \mu)^2 + y^2}}. \tag{3.12}$$

These two values will be used in the description of the numerical results.

3.4 Numerical Results

This section is devoted to explain the numerical results we obtain for the transfer methodology explained in the previous section. We have done a systematic exploration of the transfer problem varying those parameters that affect the total velocity cost $\Delta v_{tot} = \Delta v_1 + \Delta v_2$. These parameters are:

- The $z-$amplitude of the departing halo orbit.
- The point P_1 at which the departure manoeuvre is performed.
- The inclination i_1 to be achieved by $\Delta \boldsymbol{v}_1$ at P_1.
- The orbit selected for the manifold leg of the unstable manifold.
- The altitude of the circular polar orbit.

We remark that another parameter that could be considered is the change of inclination Δi_2 to be achieved by $\Delta \boldsymbol{v}_2$ at P_2, which is also the angle between the incoming and outgoing velocities at P_2; nevertheless the value of Δi_2 is not independent of the above parameters since it is mainly determined by i_1.

3.4.1 Departing Halo Orbits and Their Invariant Manifolds

We have used as departing LPOs the halo orbits of class I around the libration point L_1, whose sense of revolution about L_1 is clockwise when viewed from the Moon (see Ref. [21]).

The computation of these orbits has been done by means the Lindstedt–Poincaré (LP) procedure (see Refs. [9, 10]), and the initial conditions obtained have been refined using Newton's method. Some sample of the orbits obtained is shown in Fig. 3.5, and the corresponding initial conditions are given in Table 3.1.

The computation of the stable and unstable manifolds of the halo orbits has been done using their linear approximation given by eigenvalues and eigenvectors of the

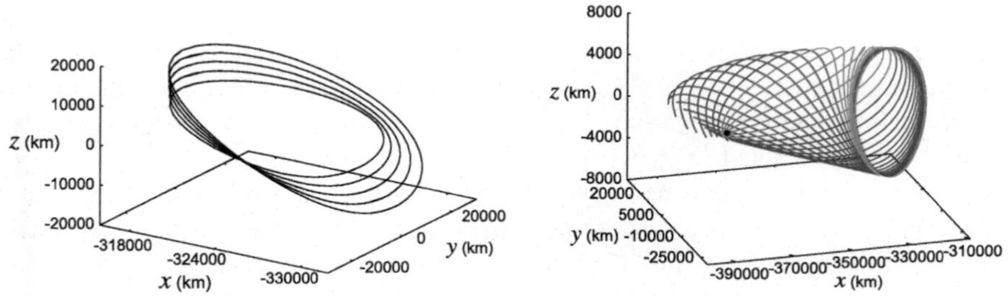

Fig. 3.5 Left: Halo orbits of class I around the L_1 point of the Earth-Moon system, associated to the initial conditions given in Table 3.1. Right: Orbits of the unstable manifold, of the halo orbit with $z-$amplitude equal to 5620.5 km, propagated until their first perilune. The orbits in red collide with the surface of the Moon

Table 3.1 Initial conditions $(x, 0, z, 0, \dot{y}, 0)$ and periods T of several halo orbits around the L_1 point of the Earth-Moon system

x (km)	z (km)	$\dot{y}$ (m/s)	T (day)
−316507.4	5620.5	−132.8	11.9
−316508.9	8298.8	−136.8	11.9
−316519.0	10783.1	−141.6	12.0
−316541.2	13118.0	−147.0	12.0
−316577.7	15343.8	−152.6	12.0

monodromy matrix. If $\boldsymbol{x}_h$ is the initial condition of the halo orbit and $\boldsymbol{w}^u$ is a unitary eigenvector along the unstable direction at this point, then

$$\boldsymbol{x}_0^u = \boldsymbol{x}_h \pm \varepsilon \boldsymbol{w}^u,$$

for a given displacement ε (typically $\varepsilon = 10^{-6}$), gives the initial condition of two orbits of the unstable manifold, one in each branch of the manifold. We use the one that approaches the Moon. For the globalisation of the manifold, we take the images of the initial condition $\boldsymbol{x}_h$ under the flow ϕ_t associated to the differential equations, $\phi_t(\boldsymbol{x}_h) = x_h(t)$, together with those of $\boldsymbol{w}^u$ under the differential of the flow, $D\phi_t(\boldsymbol{w}^u) = \boldsymbol{w}^u(t)$. In this way we can obtain a set of orbits of the unstable manifold, equally space in time along the halo orbit, whose initial conditions are,

$$\boldsymbol{x}_t^u = \boldsymbol{x}_h(t) + \varepsilon\, \boldsymbol{w}^u(t), \qquad t = t_0 + (j-1)(T/n)), \quad j = 1, ..., n, \qquad (3.13)$$

where T is the period of the halo orbit. Some of the orbits of the unstable manifold associated to the first halo orbit in Table 3.1 are shown in Fig. 3.5.

3.4.2 Selection of the Inclination i_1

It is clear that if we want to target a polar orbit, the inclination i_2 achieved by the insertion manoeuvre must be always 90°, but the selection of the suitable inclination i_1 for the first manoeuvre has more freedom. In this section we study how the value of i_1 affects the whole transfer cost.

In the first exploration we fix the initial departing halo orbit (the first one in Table 3.1), an orbit of its unstable manifold, and the point on the orbit where we do the departure manoeuvre Δv_1. This one corresponds to the first perilune on the orbit. At this point we vary the value of i_1 within $[i_x,\ 180° - i_x]$, where i_x is the angle between $\boldsymbol{r}_m$ and the (x', y') plane. For each value of i_1 we compute the transfer costs Δv_1, Δv_2 and $\Delta v_{tot} = \Delta v_1 + \Delta v_2$.

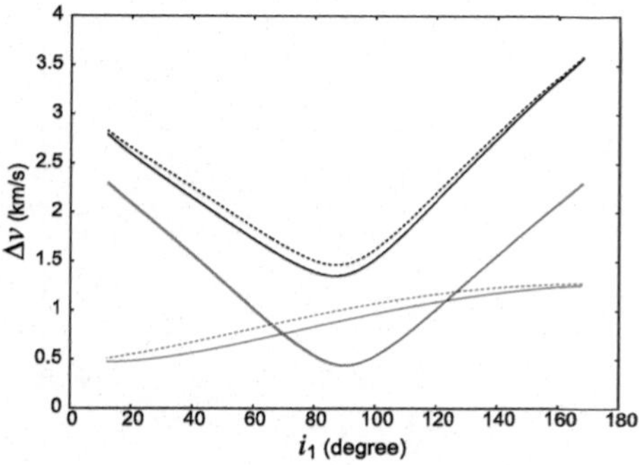

Fig. 3.6 General behaviour of the transfer costs Δv_1 (red), Δv_2 (blue) and $\Delta v_{tot} = \Delta v_1 + \Delta v_2$ (black) as a function of the inclination i_1. The point P_1 is at a distance of 1, 0202.31 km, the radius of the polar orbit is of 2037.1 km from the Moon, and $i_x = 11.12°$, so $i_1 \in [11.12°,\ 168.88°]$

Figure 3.6 shows the behaviour of Δv_1, Δv_2 and Δv_{tot} as a function of the inclination i_1 when the target orbit is a lunar polar orbit ($i_2 = 90°$). The patterns shown in this figure are almost identical to the results when other orbits of the unstable manifold are used. There are several things to be mentioned:

1. The value of Δv_1 increases with i_1, and has an almost constant small slope. Because of this fact, it is better (in terms of Δv_{tot}) to perform the change of inclination with the departure manoeuvre.
2. The value of Δv_2 has a minimum at $i_1 \approx 90°$ and the curve $(i_1, \Delta v_2)$ is almost symmetric with respect to this minimum. This is consistent with the fact that if $\Delta \boldsymbol{v}_1$ has already done 'almost' all the total change of inclination, the main role of Δv_2 is to do the insertion into the polar orbit.
3. In this example, the distance r_m from P_1 to the Moon is not too large. When i_1 is close to 90°, the value of Δv_2 is smaller than Δv_1. We will see later, this is not true if r_m is above a certain value.
4. When the inclination of the final orbit is $i_2 \neq 90°$, the symmetry and the minimum properties mentioned in the above item hold for i_1 equal to i_2.
5. The two costs associated with the departure manoeuvre Δv_1 (red curves in Fig. 3.6) are very similar, and correspond to the two possible solutions $\boldsymbol{v}_{i-t}^{1,2}$ mentioned in Sect. 3.3.2. The one with larger values of Δv_1 corresponds to the larger values of the angle ϕ_1.
6. For each value of Δv_1 there are two possible values of Δv_2. As was explained in Sect. 3.3.4, one of them is always discarded. As a result, instead of having four different $(i_1, \Delta v_2)$ curves we have only two. In Fig. 3.6 both curves almost coincide.
7. The near coincidence of the two $(i_1, \Delta v_2)$ curves is because $\Delta \boldsymbol{v}_1$ is performed at the perilune and the two transfer orbits are almost 'symmetric'. This is shown in Fig. 3.7 where the latitude of the insertion point P_2 is displayed as a function of i_1 for the two possible velocities. If the first manoeuvre is done at another point the symmetry is broken and the two costs are different.

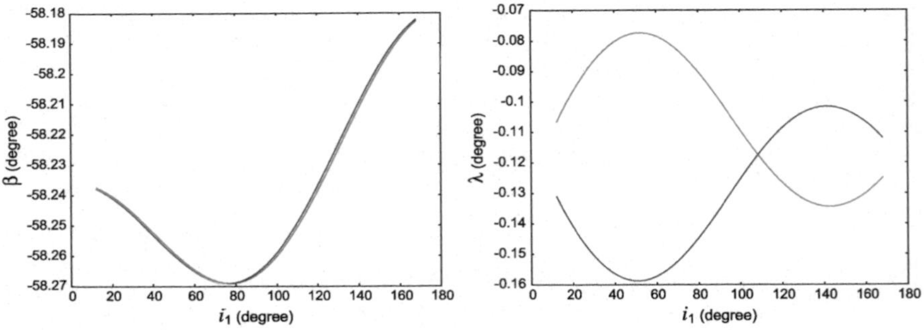

Fig. 3.7 Location of the longitude β (left) and latitude λ (right) of the point P_2 on the sphere S, as a function of i_1 and for the two possible velocities $\boldsymbol{v}_{i-t}^{1,2}$

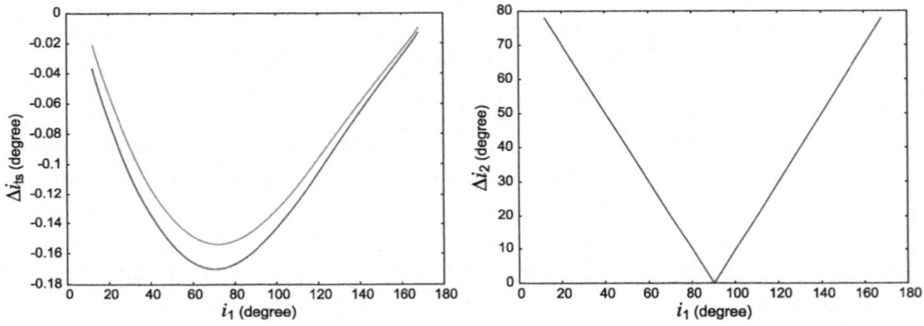

Fig. 3.8 Variation of the inclination along the transfer ellipse with respect to i_1 (left), and change of inclination that must be done by the insertion manoeuvre $\Delta \boldsymbol{v}_2$ (right). The two curves of the left plot correspond to the two velocities $\boldsymbol{v}_{i-t}^{1,2}$

8. As it can be seen in Fig. 3.7, the longitude and latitude of the point P_2 are almost constant (variations less than $0.1°$), so all the transfer ellipses are almost identical except for an orbit plane rotation along the P_1–P_2 line. Recall that by construction, the semi-major axis is approximately the segment P_1–P_2 (see Sect. 3.3).

Since the initial velocity $\boldsymbol{v}_{i-t}^{sid}$ of the transfer ellipse is computed in the 2BP model and the integration of the transfer leg is done in the CR3BP framework, the osculating inclination of the transfer ellipse after Δt_2 time units is slightly different (less than $0.2°$) from the inclination at the departing point of the ellipse. This means that, even if we set $i_1 = 90°$, $\boldsymbol{v}_{f-t}^{sid}$ is not going to be on the polar plane. The second manoeuvre also takes care of this variation. Note that when Δt_2 is large, this orbit inclination difference also becomes large. This happens when we perform the departure manoeuvre far from the Moon. Figure 3.8 shows the variation of the inclination along the transfer ellipse with respect to i_1, together with the change of inclination that must be performed by the second manoeuvre $\Delta \boldsymbol{v}_2$.

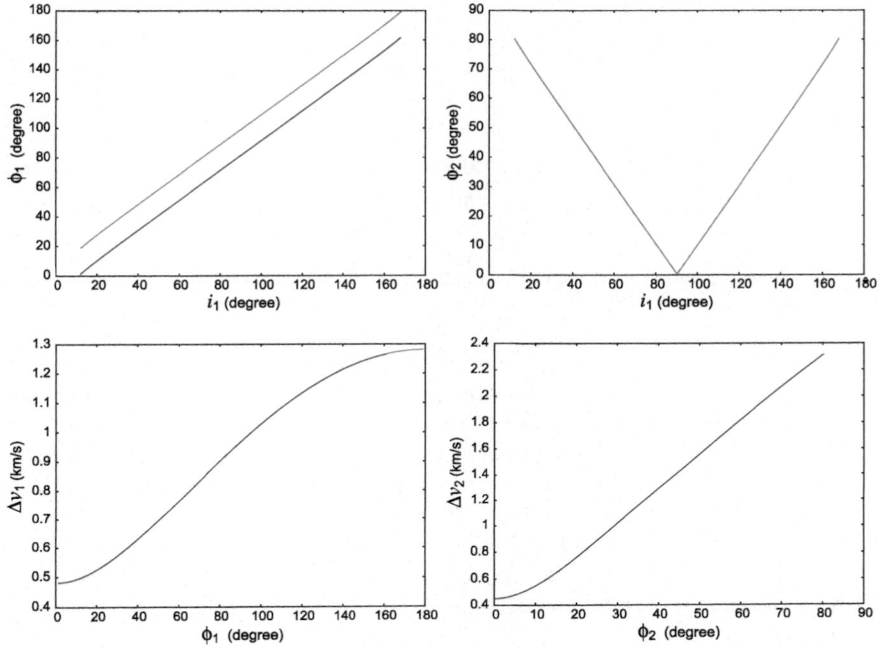

Fig. 3.9 The top figures display the relation between i_1 and $\phi_1^{1,2}$ (left) and $\phi_2^{1,2}$ (right). The bottom figures show the relation of the Δv_1 and ϕ_1 (left), and Δv_2 and ϕ_2 (right)

3.4.3 *Role of the Angles Between the Arrival and Departure Velocities at P_1 and P_2*

At the departure point P_1, the angles $\phi_1^{1,2}$ between $\boldsymbol{v}_m$ and the two departure velocities $\boldsymbol{v}_{i-t}^{1,2}$ are closely related to the value of Δv_1. Once P_1 is selected, the inclination i_m of the osculating ellipse is fixed. Furthermore, if P_1 is at the apocentre then ϕ_1 is (approximately) equal to Δi_1, in this case $\phi_1^{1,2} \approx \Delta i_1 = i_1 - i_m$, with $i_1 \in [i_x, 180° - i_x]$. Since i_m is fixed, the relation between i_1 and $\phi_1^{1,2}$ is linear, as shown by Fig. 3.9. Combining this linear relation with the one between i_1 and Δv_1 (Fig. 3.6) we get the dependence between Δv_1 and $\phi_1^{1,2}$, which is displayed in Fig. 3.9. Here the two curves overlap, which demonstrates the dependence of Δv_1 on the angle ϕ_1 when the departure point P_1 is close to the Moon.

The constant difference between the two values of ϕ_1 shown in Fig. 3.9 is due to the fact that P_1 is chosen to be at the first perilune. The incoming velocity $\boldsymbol{v}_m$ of P_1 is perpendicular to the position vector $\boldsymbol{r}_m$. According to Eq. (3.4), the two possible outgoing velocities $\boldsymbol{v}_{i-t}^{1,2}$ are also perpendicular to $\boldsymbol{r}_m$ and, as a consequence, the three velocities $\boldsymbol{v}_m$ and $\boldsymbol{v}_{i-t}^{1,2}$ are coplanar, so $|\phi_1^1 - \phi_1^2| = 2i_m \approx 20°$. This fact

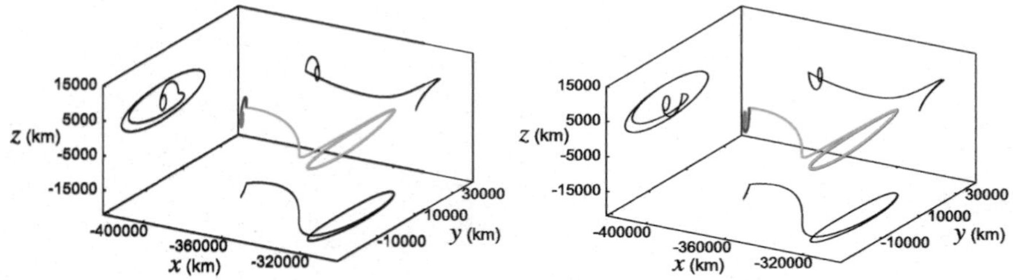

Fig. 3.10 Transfer trajectories obtained using the two values of $\mathbf{v}_{i-t}^{1,2}$, when P_1 is taken to be the first perilune and $i_1 = i_2 = 90°$

only happens if P_1 is at the peri/apolune and, at the same time, P_1 is assumed to be at the apo/pericentre of the transfer ellipse.

The angle ϕ_2 is defined in a similar way as ϕ_1 but at the point P_2. At this point the modulus of $\mathbf{v}_c^{sid}$ is given by Eq. (3.9) and the procedure makes sure that the position vector at P_2 is perpendicular to $\mathbf{v}_c^{sid}$. At P_2 the velocity along the circular polar orbit is also perpendicular to the position vector $\mathbf{r}_{f-t}$, so the angle between the arrival plane and the target plane is exactly the angle between the two velocities. If the departure manoeuvre accounts for the final change of inclination, then the angle ϕ_2 between both velocities is small.

As we can see in Fig. 3.9, the minimum insertion manoeuvre Δv_2 takes place for $i_1 = 90°$, since we are aiming at a lunar polar orbit ($i_2 = 90°$). The trajectory profiles with minimum transfer costs ($i_1 = 90°$), obtained with the two possible values of $\mathbf{v}_{i-t}$, are displayed in Fig. 3.10. The two Hohmann legs of these transfers are symmetric with respect to the (x', y') plane because the two $\mathbf{v}_{i-t}$ are aligned but point to opposite directions. The left plot corresponds to the case that needs less Δv_{tot}, which is associated with the smaller ϕ_1. It also corresponds to $\mathbf{v}_{i-t}$ with positive $z'-$component, leading to a transfer leg above the $(x' - y')$ plane.

The above analysis shows the strong dependence of the transfer costs Δv_1 and Δv_2 on ϕ_1 and ϕ_2, respectively, when P_1 is close to the Moon. It suggests that in order to reduce the total transfer cost, the appropriate strategy is to select the desired inclination i_1 to make ϕ_2 as small as possible. In other words, we should accomplish the change of inclination manoeuvre mainly at the departure point P_1. To this end we could set i_1 to be the final desired inclination.In this case study, the polar orbit is chosen as the targeting orbit, so , so we have $i_1 = i_2 = 90°$. This conclusion is the one to be expected considering a two-body approximation to the problem, for which the approximation is not so good if the point P_1 is far away from the primary.

From now on, we will only consider the transfer that requires a smaller Δv_{tot}.

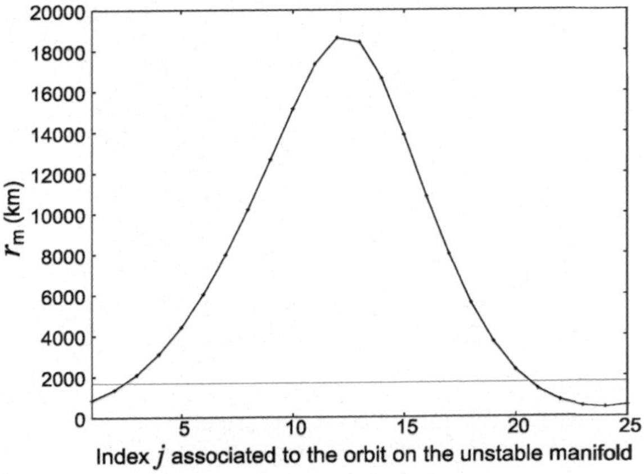

Fig. 3.11 Distance from P_1 to the Moon as a function of the index j associated to the family of unstable orbits. The horizontal line indicates the value of the radius of the Moon, $r_{Moon} = 1737.1$ km

3.4.4 The Role of the Orbit on the Unstable Manifold with Δv

In this section we show how the value of Δv varies when we consider different orbits of the unstable manifold as 'manifold legs'.

For the halo orbits under consideration, using Eq. (3.13), we have computed several unstable orbits on its unstable invariant manifolds, which have been parametrised by an integer $j = 1, ..., 25$. The first perilune of each orbit is selected as the departure manoeuvre point P_1, and we have fixed $i_1 = i_2 = 90°$. Figure 3.11 shows the distance from P_1 to the Moon as a function of the index j. From this figure it follows that some orbits of the unstable manifold ($j = 1, 2, 21, 22, 23, 24, 25$) collide the Moon's surface before they reach the first perilune. For these orbits we only perform the insertion manoeuvre $\Delta \boldsymbol{v}_2$ when the spacecraft reaches the surface S. We call them *direct insertion transfers*, to be distinguished from the usual *Hohmann-type transfers*.

Figure 3.12 shows Δv_1, Δv_2 and Δv_{tot} as a function of the parameter j, and as a function of the distance r_m from P_1 to the Moon. The three black segments of the left plot give the value of Δv_{tot} for these transfers. From this figure it follows that the *direct insertion transfers* are not a good option in terms of velocity costs, which are all greater than 2.5 km/s, while for the Hohmann-type ones the cost is much cheaper, less than 2.4 km/s. A similar fact, about higher costs associated to direct insertion transfers, was also detected and commented in Ref. [2]. The minimum cost is of 1.149 km/s ($\Delta v_1 = 0.619$ km/s, $\Delta v_2 = 0.530$ km/s), corresponding to a value of $r_m = 18407.554$ km, which is the largest first perilune distance among all the non-collision orbits of the unstable manifold. This fact will be taken into account later. From now on we focus on the *Hohmann-type transfers*.

As it was already noted, the distance r_m plays an important role in the transfer cost. With larger r_m we get smaller Δv_1 and Δv_{tot}, while the insertion manoeuvre Δv_2 shows exactly the opposite change trend. The departure manoeuvre Δv_1 is the

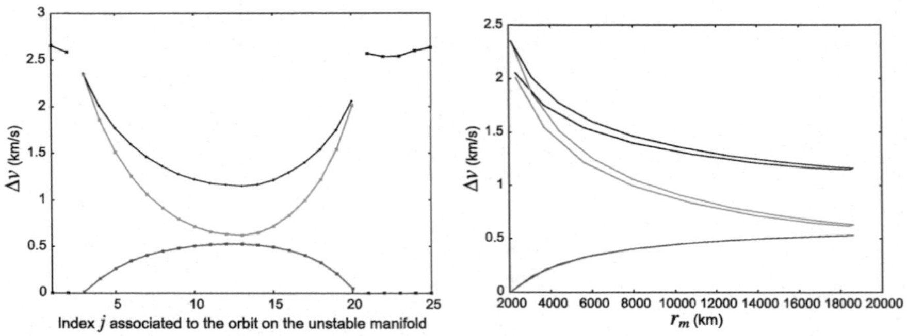

Fig. 3.12 Transfer costs as a function of the parameter j, associated to the family of orbits of the unstable manifold (left), and as a function of the distance from P_1 to the Moon (right). In both plots Δv_1 is in red, Δv_2 in blue and Δv_{tot} in black

dominant part of the total transfer cost when P_1 is not too far away from the Moon ($r_m \leq 20{,}000\,\text{km}$).

3.4.5 Varying P_1 Along the Manifold Leg

Next we want to study the possibility of performing the departure manoeuvre at a point different from the first perilune of the manifold leg. With this purpose we have analysed the orbits of the unstable manifold that do not collide with the Moon ($j \in [3, 20]$). A first exploration allows us to classify them into two kinds: those that remain captured by the Moon for at least one revolution ($j \in [15, 20]$) and those that escape before performing one revolution ($j \in [3, 14]$), which means that there exists no apolune, or the first apolune is beyond a certain distance.

We have studied the transfers using both kinds of orbits and varying the point P_1. The departing point P_1 has been determined integrating forward and backward in time from the perilune point, with time steps of 1 hour. To use the results of the 2BP approximation as the initial guess, we set an upper bound of the distance to the Moon r_m as 55,000 km. Figure 3.13 (left) shows the behaviour of r_m for the capture and escape orbits. The minima of this figure correspond to the perilunes and the maxima to the apolunes. In the left plot one can see that the orbit of the unstable manifold performs four revolutions around the Moon, while in the right plot after the first perilune the orbit 'escapes' from the Moon's neighbourhood.

For the simulations we have fixed $i_1 = i_2 = 90°$ and we show only the results corresponding to one manifold leg of each class. The qualitative behaviour is the same for all the legs of the same class and the quantitative differences are small.

For the two cases under consideration, Fig. 3.14 shows the costs Δv_1, Δv_2 and Δv_{tot} as a function of the time Δt_1 spent since the departure from the halo orbit. In both cases the Δv_1 cost is strongly related to the r_m distance, in the sense that the

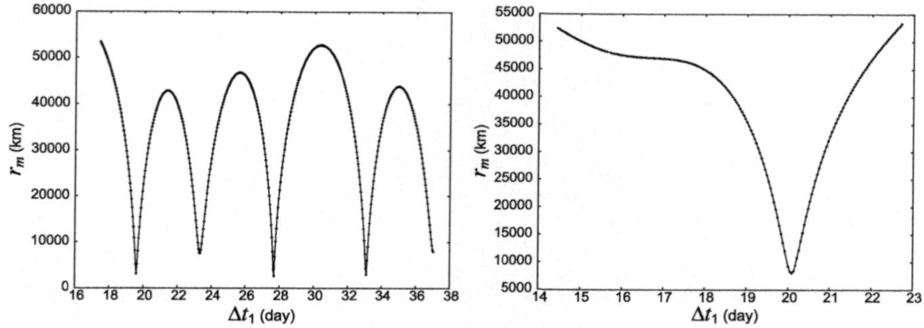

Fig. 3.13 Behaviour of r_m for the capture (left) and escape (right) orbits as a function of the time Δt_1 spent since the departure from the halo orbit

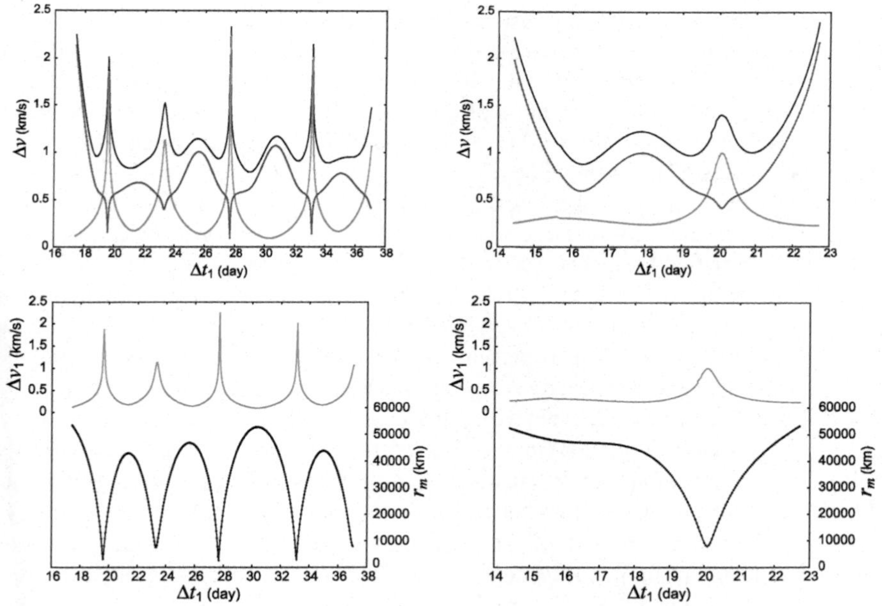

Fig. 3.14 For the capture case (top left) and the escape case (top right), behaviour of Δv_1 (red), Δv_2 (blue) and Δv_{tot} (black) in front of Δt_1. For both cases, the two bottom figures show the behaviour of Δv_1 (red) and the distance from P_1 to the Moon (black) as a function of Δt_1

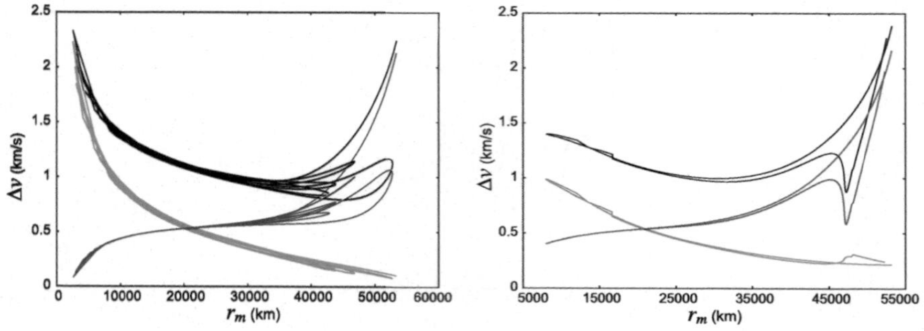

Fig. 3.15 For the capture case (left) and the escape case (right), behaviour of the costs Δv_1 (red), Δv_2 (blue) and Δv_{tot} (black) versus the distance r_m

minimum values of the cost are associated to the maxima (apolunes) of the distance to the Moon, as it is clearly shown also in Fig. 3.14. For the capture case it can also be noted that for small values of Δt_1, the first minima of Δv_{tot} (which, in fact, is a global minima) corresponds to the minima of Δv_1, and so, to the apolunes. We will make use of this fact in what follows. The exceptions to this rule (for instance $\Delta t_1 \approx 29$ days) are associated to larger values of r_m, in which case the Hohmann leg computed in a 2BP model is not a good enough approximation of the CR3BP orbit used for the transfer. In this case, Δv_2 is large since it must also account for a correction of the inclination specified by i_1 that, due to the approximation, is not achieved by Δv_1.

Looking at Fig. 3.15 we can say that when $r_m < 20{,}000$ km, the first manoeuvre Δv_1 is the main part of the total Δv, while for $r_m > 20{,}000$ km , Δv_2 becomes larger than the first manoeuvre. This is because when the departure point P_1 is far from the Moon, a small change in velocity $\boldsymbol{v}_{i-t}$ leads to a big change in the transfer trajectory. Note also that, in this figure, when we integrate backwards in time for $\Delta t_1 < 19.5$ day, and as r_m gets close to the maximum value 55,000 km , Δv_2 increases dramatically while Δv_1 decreases smoothly, so we do not consider P_1 before the first perilune. Moreover, after the intersection between the Δv_1 and Δv_2 curves, the variations of both velocities with r_m are smooth up to a certain value ($\approx$45,000 km) and afterwards they become much sharper. As a consequence, the suitable location of the departure point P_1 should be after the first perilune and with a distance less than 45,000 km away from the Moon. The first apolune seems to be a good location.

3.4.6 Setting P_1 at the First Apolune

In this section we explore the situation in which, according to the results of the preceding section, the departure point P_1 is taken at the first apolune of the manifold

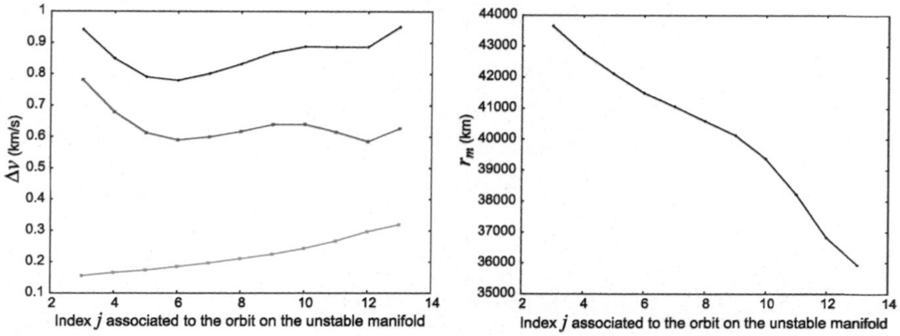

Fig. 3.16 Behaviour of the transfer costs (left) and distance r_m (right) for the captured orbits of the unstable manifold of a halo orbit

leg. In this way we avoid unnecessary revolutions around the Moon that, in principle, do not guarantee a decrease of the total Δv_{tot}.

Keeping fixed both the departing halo orbit and the target lunar polar orbit, we consider all the captured orbits of the unstable manifold that do not collide with the Moon and that perform at least one revolution around it ($j \in [3, 13]$). For all this range of orbits, Fig. 3.16 shows the results that have been obtained.

As it was already observed, when r_m is greater than 37,000 km, Δv_2 is the dominant part of Δv_{tot}. According to Fig. 3.15, when r_m is close to the maximum value 55,000 km, Δv_1 increases dramatically while Δv_2 decreases smoothly. It has also been observed that the value of Δv_2 is also strongly influenced by the angle ϕ_2, in the sense that the minima of Δv_2 correspond to the minima of ϕ_2. In terms of the total transfer cost, the apolune case requires less than 0.9 km/s with a minimum of 0.77977 km/s, while in the perilune case the minimum is of 1.14915 km/s.

3.4.7 *Changing the Sizes of the Departing and the Target Orbits*

In this section we study how the size of the initial halo orbit and the radius of the lunar polar orbit affect the transfer cost. For both explorations we keep $i_1 = i_2 = 90°$ as in the preceding sections.

For each halo in Table 3.1, we have computed 25 orbits of their unstable manifold, and along each orbit we have performed the departure manoeuvre at both the first apolune and the first perilune.

Figure 3.17 shows the costs of the transfer as a function of the index j of the orbits of the manifold in the two cases considered, and Table 3.2 gives the minimum values of Δv_{tot}. Clearly, the pattern behaviours for the $\Delta v's$ are similar to the ones we have already shown for a fixed halo amplitude. In the apolune case, the $z-$amplitude barely

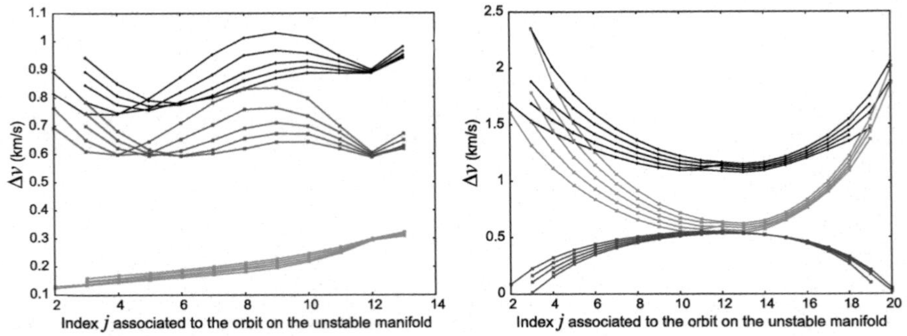

Fig. 3.17 Transfer costs in the apolune case (left) and perilune case (right) computed using the orbits on the unstable manifolds of the halo orbits given in Table 3.1. In the left plot the $z-$amplitude increases from top to bottom along the families of curves while in the right plot the sense is reversed

Table 3.2 Minimum transfer costs in unit of km/s for the halo orbits explored

z-amplitude (km)	Apolune Δv_{tot}	Perilune Δv_{tot}
5620.4505	0.77977	1.14915
8298.7933	0.77073	1.13316
10783.0642	0.75623	1.11611
13118.0261	0.74571	1.09389
15343.8285	0.74168	1.07513

plays a role, with merely 30 m/s decrease in Δv_{tot}, and the minimum transfer cost happens with the smallest ϕ_2. In the perilune case, the minimum transfer cost happens at the point with the biggest r_m, but a change in the z-amplitude of about 10,000 km only leads to about 70 m/s decrease in Δv_{tot} (see Table 3.2). In the figure, the curves of the apolune case, when seen from top to bottom, correspond to increasing values of the halo-amplitude. In the perilune case, the 'optimal' orbit of the unstable manifold is always the one associated to the upper point of the halo orbit ($j = 13$).

The value of the j parameter of the optimal apolune case, which varies between 4 and 6, indicates that the 'optimal' orbit of the unstable manifold departs close to the right hand corner of the halo orbit, as seen from the Moon.

To complete the analysis, and keeping fixed the departing halo orbit as the one with $z-$amplitude equal to 15343.83 km , we have explored how the size of the lunar polar orbit affects the $\Delta v's$. The results are shown in Fig. 3.18 and Table 3.3, from which it can be seen that the transfer cost decreases with the altitude of the final polar orbit. Note that in this case: P_1 is unchanged, and so the unstable manifold is unchanged, as a consequence, Δv_1 is barely affected by the target polar orbit. P_2 varies with the altitude of the polar orbit, and so does the velocity along the manifold leg at P_2, this explains the results shown in Fig. 3.18.

Finally we have also considered halo orbits of Class II, which are symmetric with respect to the $z = 0$ plane of Class I. Of course the results are the same as for Class I.

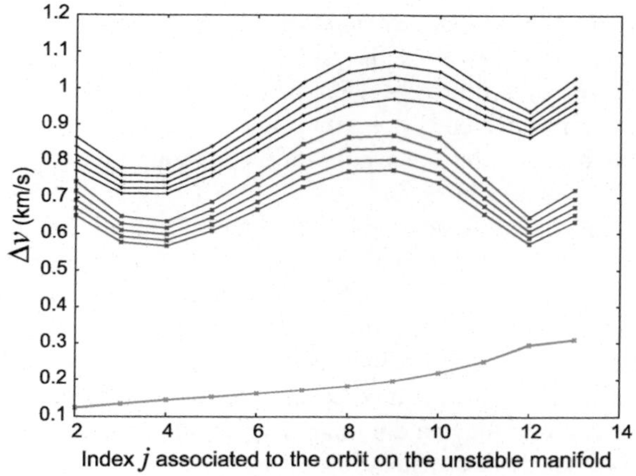

Fig. 3.18 Transfer costs for a fixed departing halo orbit and 5 lunar polar orbits with different altitudes. The curves of both plots correspond, from top to bottom, to increasing values of r_{polar}

Table 3.3 Minimum Δv_{tot} for a fixed departing halo orbit, $j = 4$, and target polar orbits of different altitudes

Altitude (km)	100	200	300	400	500
Δv_{tot} (km/s)	0.77760	0.75894	0.74168	0.72567	0.71076

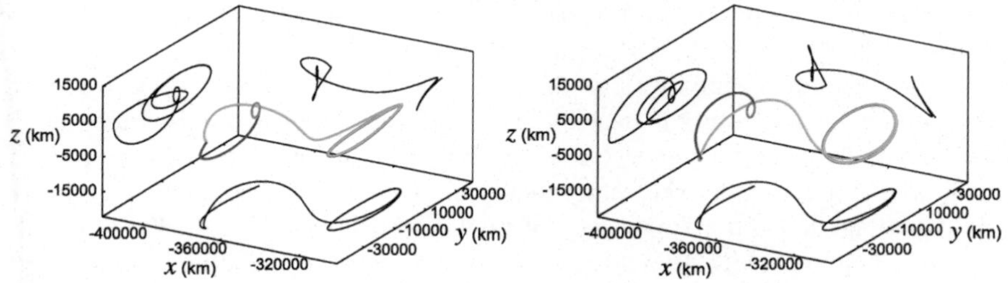

Fig. 3.19 Transfer trajectory profiles for the minimum cost cases in the apolune case for Class I halo orbit (left) and Class II halo orbit (right)

Figure 3.19 shows the transfer trajectories with the minimum cost obtained for both families where the symmetry with respect to $z = 0$ plane between the two families is clearly seen.

3.5 Conclusions

In this chapter we have done an exploration of the possible parameters that may affect the transfer cost from LPOs to lunar orbits. In the Earth–Moon CR3BP we have developed a general procedure to perform transfers from a LPO to a circular polar orbit around the Moon. A key point of the procedure has been the use of the unstable manifold associated with the LPO, which implies that at most only two transfer manoeuvres are required: a departure and an insertion manoeuvre. All the numerical simulations have been done for transfers between halo orbits around the L_1 libration point and circular lunar polar orbits, although the procedure is general for many other classes of orbits.

One of the objectives of the study is to analyse the role of the different free parameters of the problem in the total transfer cost. The two components of the total transfer cost $\Delta v_{tot} = \Delta v_1 + \Delta v_2$ have been analysed independently.

Among the parameters, the most important ones are: the distance from the Moon r_m at the departure manoeuvre, the inclination i_1 to be achieved by this manoeuvre, or the change of inclination Δi_2 to be performed by the insertion manoeuvre Δv_2. In general, no parameter plays a dominant role in all the situations that have been considered. However, there are some clear patterns. When departing from a point on the manifold close to the Moon, say $r_m < 20{,}000$ km, the departure manoeuvre Δv_1 dominates the cost and decreases when r_m increases. In this situation Δv_{tot} is always greater than 1 km/s. When $r_m > 20{,}000$ km, Δv_2 becomes the dominant part of Δv_{tot}. When r_m is greater than 45,000 km, Δv_2 increases dramatically. The value of Δv_2 is strongly correlated with the change of inclination that the second manoeuvre must perform which, at the same time, is strongly influenced by r_m if $i_1 = i_2$. In this situation Δv_{tot} can be smaller than 0.8 km/s.

It must also be noted that a direct transfer from the departing halo orbit to the lunar polar orbit is not a good option, since it belongs to the case in which the departure point along the manifold leg is chosen before the first perilune. In this case, the transfer cost Δv_{tot}, is always greater than 2.5 km/s.

As general conclusions we can say that, in order to have low values of Δv_{tot}: the departure point along the unstable manifold must be far from the Moon ($r_m >$ 20,000 km), but not too far away ($r_m <$ 45,000 km). In order to make Δv_2 as small as possible, the inclination to be achieved by the first manoeuvre must be set equal to the target final inclination (in the case of a polar orbit, then $i_1 = 90°$). The size of the departing halo orbit has little influence on the total cost. If the altitude of the target polar orbit increases, then the insertion manoeuvre Δv_2 decreases. Considering that the flight time should not be too long, the departure manoeuvre should be done at the first apolune of the unstable invariant manifold of the LPO.

References

1. A Standardised Lunar Coordinate System for the Lunar Reconnaissance Orbiter and Lunar Datasets, LRO Project and LGCWG White Paper, Version 5. NASA (2008), http://lunar.gsfc.nasa.gov/library/LunCoordWhitePaper-10-08.pdf
2. E.M. Alessi, G. Gómez, J.J. Masdemont, Two-manoeuvres transfers between LEOs and Lissajous orbits in the Earth-Moon system. Adv. Space Res. **45**(10), 1276–1291 (2010)
3. R.L. Anderson, J.S. Parker, Comparison of low-energy lunar transfer trajectories to invariant manifolds. Celest. Mech. Dyn. Astron. **115**(3), 311–331 (2013)

4. B.A. Archinal et al., Report of the IAU working group on cartographic coordinates and rotational elements: 2009. Celest. Mech. Dyn. Astron. **109**(2), 101–135 (2011)
5. G.H. Born, J.S. Parker, Direct halo lunar transfers. J. Astronaut. Sci. **56**(4), 441–476 (2008)
6. J. Cao, S. Hu, Y. Huang, Orbit determination and analysis for Chang'E-2 extended mission. Geomat. Inf. Sci. Wuhan Univ. **38**(9), 1029–1033 (2013)
7. P.R. Escobal, *Methods of Astrodynamics* (Wiley, New Jersey, 1969)
8. D.C. Folta, M. Woodard, K.C. Howell, C. Patterson, W. Schlei, Applications of multi-body dynamical environments: the ARTEMIS transfer trajectory design. Acta Astronaut. **73**, 237–249 (2012)
9. G. Gómez, J.-M. Mondelo, The Dynamics Around the Collinear Equilibrium Points of the RTBP. Phys. D: Nonlinear Phenom. **157**(4), 283–321 (2001)
10. G. Gómez, J. Llibre, R. Martínez, C. Simó, *Dynamics and Mission Design Near Libration Points*, Vol. I Fundamentals: The Case of Collinear Libration Points (World Scientific, Singapore, 2001)
11. G. Gómez, J.J. Masdemont, J.-M. Mondelo, Libration point orbits: a survey from the dynamical point of view, in *Proceedings of the Conference Libration Point Orbits and Applications* ed. by G. Gómez, M.W. Lo, J.J. Masdemont. (World Scientific, 2003), pp. 311–372
12. W.S. Koon, M.W. Lo, J.E. Marsden, S.D. Ross, Heteroclinic Connections Between Periodic Orbits and Resonance Transitions in Celestial Mechanics. Chaos **10**, 427–469 (2000)
13. W.S. Koon, M.W. Lo, J.E. Marsden, S.D. Ross, Low Energy Transfer to the Moon. Celest. Mech. Dyn. Astron. **81**, 63–73 (2001)
14. W.S. Koon, M.W. Lo, J.E. Marsden, S.D. Ross, *Dynamical Systems, the Three-Body Problem and Space Mission Design* (Marsden Books, 2011). ISBN 978-0-615-24095-4
15. L. Liu, Y. Liu, J. Cao, S. Hu, G. Tang, J. Xie, CHANG'E-2 lunar escape maneuvers to the sun earth L_2 libration point mission. Acta Astronautica **93**, 390–399 (2014)
16. J.K. Miller, Lunar transfer trajectory design and the four body problem, in *13th AAS/AIAA Space Flight Mechanics Meeting*, Paper No. AAS 03-144 (2002)
17. J.S. Parker, R.L. Anderson, *Low-Energy Lunar Trajectory Design*, Deep Space Communications and Navigation Series (DESCANSO) (2013)
18. H. Peng, Y. Wang, J.J. Masdemont, G. Gómez, Design and analysis of transfers from Lunar polar orbits to sun-earth libration point orbit. (In preparation)
19. Y. Ren, J. Shan, Low-energy Lunar transfers using spatial transit orbits. Commun. Nonlinear Sci. Numer. Simul. **19**(3), 554–569 (2014)
20. Á. Rincón, P. Rojo, E. Lacruz, G. Abellán, S. Díaz, On Non-coplanar Hohmann transfer using angles as parameters. Astrophys. Space Sci. **359**, 1–6 (2015)
21. C.E. Roberts, The SOHO mission L_1 halo orbit recovery from the attitude control anomalies of 1998, in *Proceedings of the Conference Libration Point Orbits and Applications*, ed. by G. Gómez, M.W. Lo, J.J. Masdemont (World Scientific, 2003), pp. 171–218
22. V. Szebehely, *Theory of Orbits* (Academic Press, New York, 1967)

Chapter 4
Lorentz Force Formation Flying in the Earth-Moon System

We analyse a dynamical scenario where a constantly charged spacecraft (follower) moves in the vicinity of another one (leader) that follows a circular Keplerian orbit around the Earth and generates a rotating magnetic dipole. The mass of the follower is assumed to be negligible when compared with the one of the leader and both spacecrafts are supposed to be in a high-Earth orbit, so the Lorentz force on the follower due to the geomagnetic field is ignored. With these assumptions, the motion of the leader is not perturbed by the follower and is only subjected to the Earth's gravitational force, while the charged follower is subject to both the gravitational force of the Earth and the Lorentz force due to the magnetic dipole of the leader.

We focus on the dynamical characteristics of the system as a function of its parameters, with special attention to the ratio of the leader's mean motion around the Earth to the rotating rate of the dipole. We study the critical points of the model and their stability, the admissible and forbidden regions of motion of the deputy using the zero velocity surfaces and the families of periodic orbits emanating from the equilibria. In the normal case we pay special attention to the periodic orbits of elliptic type and to the families of 2D tori surrounding them that are computed by means of a parameterisation method. The result is a fine catalog of orbits together with an accurate dynamical description suitable to researchers interested in potential applications of satellite formation flight using this kind of technology. Several configurations of formation flying have been designed for demonstration purpose in the future application.

4.1 Introduction

When a charged spacecraft moves in a magnetic field, it experiences a Lorentz force. This propellant-less force, which is perpendicular to both the instantaneous velocity of the spacecraft and the magnetic field, offers some advantages when compared with the traditional chemical propulsion and is a promising technology for future space missions. The magnetic dipole of the leader is supposed to be produced by

J. Yuan et al., *Low Energy Flight: Orbital Dynamics and Mission Trajectory Design*, https://doi.org/10.1007/978-981-13-6130-2_4

three concentric and orthogonal high temperature super-conducting coils (HTSC) [21], in such a way that it can point along any direction adjusting the current in the three HTSC. Here we will only consider three possible orientations: normal, radial and tangential, according to its relative position in the orbital plane of the leader.

Previous works on applications of Lorentz forces in space missions have mainly considered the effect of natural magnetic fields on a charged spacecraft, such as the geomagnetic field [10, 19, 36], or the magnetic filed of other planets [27] on the spacecraft. Some of the applications that have been studied include the determination of new synchronous orbits using the geomagnetic field [31], the use of the Lorentz force as a means for orbit control and formation flying [20, 34], and the effect of this force in a Jovian orbit insertion [2] as well as in gravity-assist manoeuvres [33]. Several control strategies using Lorentz force have also been developed for different kinds of missions [13, 29, 32].

Using artificial magnetic fields, Kong [15] introduced the idea of Electro Magnetic Formation Flight (EMFF), which uses the intersection between the electromagnetic fields of several satellites to control the configuration of the formation; Umair [35] designed a control strategy for electromagnetic satellite formations in near-Earth orbits; Kwon [16] explored the applicability of EMFF for attitude and translation control of close proximity formation flying; Porter [24] demonstrated the feasibility of electromagnetic formation flight in a micro-gravity environment.

Inspired by previous works, a new dynamical scenario was proposed by Peng [22], where a charged follower moves around a leader spacecraft which produces a rotating magnetic dipole pointing along the radial direction defined by the Earth and the leader. Peng [23] extended the previous work to study the planar periodic orbits, which are suitable for formation flying when the magnetic dipole is perpendicular to the orbital plane of the leader. Our paper aims to extend Peng's work by considering three possible orientations of the dipole: normal, radial and tangential and pursuing a fine dynamical analysis in all cases.

From the engineering point of view, there are some technical questions about the realisation of the system that must be considered. They are mainly related to the thermal control system of the HTSC, and the satellite charging of the follower. They have been studied, among others, by Kwon [17] who designed a cryogenic heat pipe to cool the HTSC, Baudouy [4] who surveyed the main cooling techniques at low temperature, and Saaj et al. [26] who showed that current technology can realise charge-to-mass ratio of the order of 10^{-6} to 10^{-3}.

Although the device required to generate a steerable electromagnetic dipole, using, for instance, three orthogonal electromagnetic coils made of superconducting wires, is a massive structure, we will assume that the gravitational interaction between the leader and the follower is negligible, at least when it is compared with the Lorentz force.

The above mentioned technical questions are out of the scope of this book. Our main attention is on the dynamical analysis of the model, providing an accurate description suitable for possible applications. The analysis and numerical simulations based on the assumption that all the proposed magnetic field or charge-to-mass ratio can be satisfied this assumption, although might be beyond the current state-of-art technology, is hopefully to come into reality in the near future.

This chapter is organised as follows: in Sect. 4.2 the dynamical model is developed, including the equations of motion and their symmetry properties; and then the equilibria are studied together with their stability behaviour, as well as the allowable regions of motion in the configuration space. Section 4.3 is devoted to explain the methods used for the computation of the symmetric and non-symmetric periodic orbits emanating from the equilibria, and the parameterisation method for the numerical computation of the 2D invariant tori. The numerical results obtained are shown in Sect. 4.3.3. Possible applications of the obtained results to formation flying which mainly take some of the periodic orbits obtained in this work as nominal trajectories, are discussed briefly in Sect. 4.4. Finally, Sect. 4.5 gives some conclusions.

4.2 Analysis of the Relative Dynamics of a Charged Spacecraft Moving Under the Influence of a Magnetic Field

4.2.1 Modelling Equations and Symmetries

4.2.1.1 Equations of Motion

As it has been mentioned, we assume that the mass of the follower is negligible when compared with the one of the leader, which is assumed to move in a circular high-Earth orbit (such as GEO), so that the geomagnetic Lorentz force on the follower is ignored. As a consequence, the follower is subject to both the gravitational force of the Earth and the Lorentz force due to the artificial magnetic dipole. As usual in the description of relative motions, we use a local-vertical-local-horizontal (LVLH) reference system, with the origin located on the leader, the positive x-axis ($\boldsymbol{e}_r$) pointing from the Earth to the leader, the z-axis ($\boldsymbol{e}_n$) parallel to the leader's angular momentum, and the y-axis ($\boldsymbol{e}_t$) completing a right-hand coordinate system (see Fig. 4.1).

In this LVLH reference system, the equations of relative motion of the follower with respect to the leader can be written as a perturbation of the Hill-Clohessy-Wiltshire (HCW) equations in the following way:

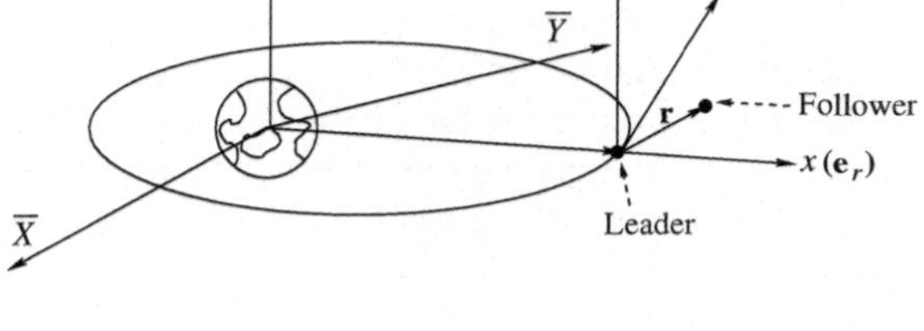

Fig. 4.1 The local-vertical-local-horizontal reference system ($\boldsymbol{e}_r, \boldsymbol{e}_n, \boldsymbol{e}_T$)

$$\begin{aligned}\ddot{x}-2n\dot{y}-3n^2x&=f_x,\\ \ddot{y}+2n\dot{x}&=f_y,\\ \ddot{z}+n^2z&=f_z,\end{aligned} \tag{4.1}$$

where n is the mean motion of the circular orbit of the leader around the Earth, and $(f_x,\ f_y,\ f_z)^T$ are the three components of the Lorentz force $\boldsymbol{f}_L$ acting on the follower.

It must be noted that if the Lorentz force $\boldsymbol{f}_L$ vanishes, then Eq. (4.1) becomes the usual HCW equations, which have been extensively studied in the literature (see, for instance, [5, 8, 25]). In this case, the dynamics of the problem is much simpler, since the HCW equations describe the relative dynamics of two Keplerian orbits, which is far different from the problem considered in this chapter where the Lorentz force has a key role.

The Lorentz force is given by:

$$\boldsymbol{f}_L=\frac{q}{m}\boldsymbol{v}_r\times\boldsymbol{B}=\frac{q}{m}\cdot(\dot{\boldsymbol{r}}-\boldsymbol{\omega}_c\times\boldsymbol{r})\times\boldsymbol{B}. \tag{4.2}$$

where, q denotes the charge of the follower, m its mass, $\boldsymbol{v}_r$ the relative velocity of the follower with respect to the rotating magnetic field, $\boldsymbol{B}$ the magnetic rotating field, $\boldsymbol{r}$ and $\dot{\boldsymbol{r}}$ the position and velocity of the follower in the LVLH reference frame, and $\boldsymbol{\omega}_c$ the dipole's rotational velocity. If ω_c is the modulus of $\boldsymbol{\omega}_c$, we can write

$$\boldsymbol{\omega}_c=\omega_c(N_x,\ N_y,\ N_z)^T, \tag{4.3}$$

where $\boldsymbol{N}=(N_x,\ N_y,\ N_z)^T$ is the unitary vector in the direction of $\boldsymbol{B}$.

The magnetic field $\boldsymbol{B}$ is defined as the curl of the potential $\boldsymbol{A}$ ($\boldsymbol{B}=\nabla\times\boldsymbol{A}$), where $\boldsymbol{A}$ is given by:

$$\boldsymbol{A}=\frac{B_0}{r^2}(\hat{\boldsymbol{N}}\times\hat{\boldsymbol{r}})=\frac{B_0}{r^2}\left[(zN_y-yN_z)\quad(xN_z-zN_x)\quad(yN_x-xN_y)\right]^T \tag{4.4}$$

where $r=\sqrt{x^2+y^2+z^2}$ is the distance from the leader to the follower, $\hat{\boldsymbol{r}}=[\hat{x}\ \hat{y}\ \hat{z}]^T=\frac{\boldsymbol{r}}{r}$ is the normalized unit vector, and B_0 is the magnetic dipole moment (unit $Wb\cdot m$) [16], which for a coil is defined by

$$B_0=\frac{\mu_0}{4\pi}n_ci_c\pi r_c^2, \tag{4.5}$$

where $\mu_0=4\pi\times10^{-7}N/A^2$ is the vacuum permeability, n_c the number of loops in the coil, i_c the current flown intensity, and r_c the radius of the coil (see Ref. [16]). Once the size of the coil (n_c and r_c) is fixed, the magnetic dipole moment is determined by the value of current i_c passing through, which, according to the material of the coil and working temperature, is limited by the current density of the coil. In our mathematical analysis, we assume that the coil can carry enough current to produce the required magnetic moment.

Using the above notation, the components of the Lorentz force $\boldsymbol{f}_L$ can be written as,

$$
\begin{aligned}
f_x &= \frac{q}{m}\frac{B_0}{r^3}\big[3(\boldsymbol{N}\cdot\hat{\boldsymbol{r}})(\dot{y}\hat{z}-\dot{z}\hat{y}) - \omega_c(xN_z - zN_x)\big(3(\boldsymbol{N}\cdot\hat{\boldsymbol{r}})\hat{z} - N_z\big) \\
&\quad + \dot{z}N_y - \dot{y}N_z + \omega_c(yN_x - xN_y)\big(3(\boldsymbol{N}\cdot\hat{\boldsymbol{r}})\hat{y} - N_y\big)\big], \\
f_y &= \frac{q}{m}\frac{B_0}{r^3}\big[3(\boldsymbol{N}\cdot\hat{\boldsymbol{r}})(\dot{z}\hat{x}-\dot{x}\hat{z}) - \omega_c(yN_x - xN_y)\big(3(\boldsymbol{N}\cdot\hat{\boldsymbol{r}})\hat{x} - N_x\big) \\
&\quad + \dot{x}N_z - \dot{z}N_x + \omega_c(zN_y - yN_z)\big(3(\boldsymbol{N}\cdot\hat{\boldsymbol{r}})\hat{z} - N_z\big)\big], \qquad (4.6) \\
f_z &= \frac{q}{m}\frac{B_0}{r^3}\big[3(\boldsymbol{N}\cdot\hat{\boldsymbol{r}})(\dot{x}\hat{y}-\dot{y}\hat{x}) - \omega_c(zN_y - yN_z)\big(3(\boldsymbol{N}\cdot\hat{\boldsymbol{r}})\hat{y} - N_y\big) \\
&\quad + \dot{y}N_x - \dot{x}N_y + \omega_c(xN_z - zN_x)\big(3(\boldsymbol{N}\cdot\hat{\boldsymbol{r}})\hat{x} - N_x\big)\big].
\end{aligned}
$$

We note that, in principle, the three orthogonal super-conducting wires can produce magnetic field in any direction, but in what follows we will only consider the following three dipole orientations:

1. **Normal case.** $\boldsymbol{N} = (0,\ 0,\ \pm 1)^T$: the dipole axis is parallel to $\boldsymbol{e}_N$.
2. **Radial case.** $\boldsymbol{N} = (\pm 1,\ 0,\ 0)^T$: the dipole axis is parallel to the radius vector $\boldsymbol{e}_R$.
3. **Tangential case.** $\boldsymbol{N} = (0,\ \pm 1,\ 0)^T$: the dipole axis is parallel to $\boldsymbol{e}_T$.

Following again Ref. [22], and in order to simplify the equations of motion, we introduce a set of non-dimensional units:

- Time unit: $\tau = n\,t$. The derivative with respect to τ will be denoted by a prime, and clearly
$$\frac{d}{d\tau} = n\frac{d}{dt}.$$
- Length unit: α such that
$$\alpha^3 = \left|\frac{q}{m}B_0\frac{\omega_c}{n^2}\right| = \left|\frac{q}{m}B_0\frac{1}{n\,\beta}\right|,$$
where the parameter β is defined by $\beta = n/\omega_c$.

4.2.2 Equations of Motion in the Normal, Radial and Tangential Cases

4.2.2.1 The Normal Case

In the normal case, $\boldsymbol{B}$ is aligned with the unit vector $\boldsymbol{N} = (0,\ 0,\ \pm 1)^T$. This means that the dipole axis is perpendicular to the orbital plane of the leader, pointing to the positive or negative $z-$axis direction according to the sign of ± 1. In this case we have,

$$f_L = \pm \frac{q}{m} \frac{B_0}{r^5} \begin{pmatrix} -(x^2 + y^2 - 2z^2)\dot{y} - 3yz\dot{z} \pm \omega_c x(x^2 + y^2 - 2z^2) \\ (x^2 + y^2 - 2z^2)\dot{x} + 3xz\dot{z} \pm \omega_c y(x^2 + y^2 - 2z^2) \\ 3z\left(y\dot{x} - x\dot{y} \pm \omega_c(x^2 + y^2)\right) \end{pmatrix}. \quad (4.7)$$

In the analysis we consider only the case $\boldsymbol{N} = (0,\ 0,\ +1)^T$, since the results for other cases can be easily obtained from this one by symmetry. Using the non-dimensional units, and defining $X = x/\alpha$, $Y = y/\alpha$, and $Z = z/\alpha$, the equations of relative motion for the follower can be written as,

$$\begin{aligned} X'' - 2Y' - 3X &= \sigma \frac{-\beta(X^2 + Y^2 - 2Z^2)Y' - 3\beta YZZ' + X(X^2 + Y^2 - 2Z^2)}{R^5}, \\ Y'' + 2X' &= \sigma \frac{\beta(X^2 + Y^2 - 2Z^2)X' + 3\beta XZZ' + Y(X^2 + Y^2 - 2Z^2)}{R^5}, \\ Z'' + Z &= \sigma \frac{3Z\left[\beta YX' - \beta XY' + (X^2 + Y^2)\right]}{R^5}, \end{aligned} \quad (4.8)$$

where σ is the sign of the charge q, and $R = \sqrt{X^2 + Y^2 + Z^2}$. It must be noted that, aside from σ, the only parameter that appears in the above equations is the angular quotient $\beta = n/\omega_c$.

The above system admits a first integral, to which we will refer as *energy*, that is given by,

$$H_N = 3X^2 - Z^2 - \sigma \frac{2(X^2 + Y^2)}{R^3} - (X'^2 + Y'^2 + Z'^2). \quad (4.9)$$

4.2.2.2 The Radial Case

In the radial case, $\boldsymbol{B}$ is aligned with the radius vector of the leader $\boldsymbol{r}$, so $\boldsymbol{N} = (\pm 1,\ 0,\ 0)^T$. Again we only consider $\boldsymbol{N} = (+1,\ 0,\ 0)^T$. Using the previous notation, the equations of relative motion in non-dimensional units can be written as,

$$\begin{aligned} X'' - 2Y' - 3X &= \sigma \frac{3X\left[\beta ZY' - \beta YZ'(Y^2 + Z^2)\right]}{R^5}, \\ Y'' + 2X' &= \sigma \frac{-\beta(Y^2 + Z^2 - 2X^2)Z' - 3\beta XZX' + Y(Y^2 + Z^2 - 2X^2)}{R^5}, \\ Z'' + Z &= \sigma \frac{-\beta(Y^2 + Z^2 - 2X^2)Y' + 3\beta XYX' + Z(Y^2 + Z^2 - 2X^2)}{R^5}. \end{aligned} \quad (4.10)$$

The energy integral is now defined by,

$$H_R = 3X^2 - Z^2 - \sigma \frac{2(Y^2 + Z^2)}{R^3} - (X'^2 + Y'^2 + Z'^2). \quad (4.11)$$

4.2.2.3 The Tangential Case

In the tangential case $\boldsymbol{N} = (0,\ \pm 1,\ 0)^T$, as in the other two cases we only consider $\boldsymbol{N} = (0,\ +1,\ 0)^T$ and the non-dimensional equations of relative motion are given by,

$$\begin{aligned} X'' - 2Y' - 3X &= \sigma\,\frac{\beta(X^2+Z^2-2Y^2)Z' + 3\beta YZY' + X(X^2+Z^2-2Y^2)}{R^5},\\ Y'' + 2X' &= \sigma\,\frac{3Y\left[\beta XZ' - \beta ZX' + (X^2+Z^2)\right]}{R^5},\\ Z'' + Z &= \sigma\,\frac{-\beta(X^2+Z^2-2Y^2)X' - 3\beta XYY' + Z(X^2+Z^2-2Y^2)}{R^5}, \end{aligned} \tag{4.12}$$

with the associated energy integral,

$$H_T = 3X^2 - Z^2 - \sigma\,\frac{2(X^2+Z^2)}{R^3} - (X'^2 + Y'^2 + Z'^2). \tag{4.13}$$

4.2.3 Symmetries

In this section we study the symmetries of the three models under consideration. For this purpose, we look for transformations which can keep the equations of motion invariant. The general pattern of the transformations considered is,

$$(t, x, y, z) \quad \longrightarrow \quad (Dt, Ax, By, Cz). \tag{4.14}$$

This is, if $\big(x(t), y(t), z(t)\big)$ is a solution of a system, such as (4.8), (4.10) or (4.12), then $\big(Ax(Dt), By(Dt), Cz(Dt)\big)$ must be also a solution of the same system.

In the normal case, the parameters A, B, C and D must fulfil the following conditions: $A^2 = 1,\ \ B^2 = 1,\ \ C^2 = 1,\ \ D = AB$. Leaving aside the trivial case $A = B = C = D = 1$, this gives rise to seven different symmetries. Three of them preserve the time orientation ($D = +1$), and the other four reverse it ($D = -1$). One of the transformations is a symmetry w.r.t. the origin ($A = B = C = -1$), three are w.r.t. a coordinate axis ($A\,B\,C = +1$), and the other three are symmetric w.r.t. a coordinate plane ($A\,B\,C = -1$). The time reversing symmetries will be used for the computation of symmetric periodic orbits. Once one of these periodic orbits is computed, using any of the three symmetries preserving the time orientation we obtain three periodic orbits with the same period ($\mathrm{III}_N, \mathrm{IV}_N, \mathrm{V}_N$). Some of them are potential candidates to be used as nominal orbits for formation flying.

In the radial and tangential cases we have $A^2 = 1,\ B^2 = 1,\ C = A,\ D = AB$ and $A^2 = 1,\ B^2 = 1,\ C = B,\ D = AB$, respectively. Both cases exhibit symmetries with respect to the origin, a coordinate axis and a coordinate plane, but the total number of feasible symmetries in both cases is only three: two reversing the time orientation and one preserving it. The values of the coefficients A, B, C and D of

Table 4.1 Values of the parameters A, B, C and D of the transformation (4.14) that leave invariant the differential equations (4.8), (4.10), and (4.12) in the normal, radial, and tangential cases, respectively

Model case	$A,\ B,\ C,\ D$	Label	Symmetry element
Normal	$1, -1, -1, -1$	I_N	X axis
	$-1,\ 1, -1, -1$	II_N	Y axis
	$-1, -1,\ 1,\ 1$	III_N	Z axis
	$-1, -1, -1,\ 1$	IV_N	Origin
	$1,\ 1, -1,\ 1$	V_N	$X-Y$ plane
	$-1,\ 1,\ 1, -1$	VI_N	$Y-Z$ plane
	$1, -1,\ 1, -1$	VII_N	$X-Z$ plane
Radial	$1, -1,\ 1, -1$	I_R	$X-Z$ plane
	$-1,\ 1, -1, -1$	II_R	Y axis
	$-1, -1, -1,\ 1$	III_R	Origin
Tangential	$-1,\ 1,\ 1, -1$	I_T	$Y-Z$ plane
	$1, -1, -1, -1$	II_T	X axis
	$-1, -1, -1,\ 1$	III_T	Origin

the invariance transformations, and the symmetry properties of the three cases are listed in Table 4.1.

4.2.4 Equilibrium Points, Stability and Zero Velocity Surfaces

4.2.5 Equilibrium Points

In this section we study the equilibrium points of the equations of motion in the normal, radial, and tangential case together with their stability. Setting $X' = Y' = Z' = 0$ and $X'' = Y'' = Z'' = 0$ in the equations of motion, and solving the associated non-linear system, we obtain the location of the equilibria that only depends on σ which is the sign of the charge q of the follower.

Table 4.2 gives the results obtained in the three cases, together with the value of the energy integral H at the equilibrium point. The points are displayed in Fig. 4.2 using black, blue and purple colours, according to their location.

We have labelled the equilibrium points according to the value of the Hamiltonian (energy level) and the direction of the dipole. For a positively charged spacecraft ($\sigma > 0$), in the normal case we have four equilibria in the $Y - Z$ plane with label 1_N, and four in the $X - Z$ plane with label 2_N. When $\sigma < 0$ there are only two equilibria, both on the $X-$ axis, with label 3_N.

In the radial case, and for $\sigma > 0$, we have two equilibria on the $Z-$axis (with label 1_R). When $\sigma < 0$, there are four equilibrium points in the $X - Y$ plane (2_R), and four in the $X - Z$ plane (3_R).

Table 4.2 Equilibrium points of the system equations (4.8), (4.10) and (4.12). In the three cases, the equilibria are also labelled according to their energy level

Label	X	Y	Z	σ	H
1_N	0	$\pm\sqrt{2}Z$	$\pm\left(\frac{2}{3\sqrt{3}}\right)^{1/3}$	$+$	-1.587401
2_N	$\pm\left(\frac{1}{12\sqrt{6}}\right)^{1/3}$	0	$\pm\sqrt{5}X$	$+$	-0.629960
3_N	$\pm\left(\frac{1}{3}\right)^{1/3}$	0	0	$-$	4.326748
1_R	0	0	± 1	$+$	-3
2_R	$\pm\left(\frac{2}{9\sqrt{3}}\right)^{1/3}$	$\pm\sqrt{2}X$	0	$-$	2.289428
3_R	$\pm\left(\frac{1}{4\sqrt{2}}\right)^{1/3}$	0	$\pm\sqrt{X}$	$-$	1.88988
1_T	0	0	± 1	$+$	-3
2_T	$\pm\left(\frac{1}{3}\right)^{1/3}$	0	0	$-$	4.326748
3_T	0	$\neq 0$	0	$\pm$	0

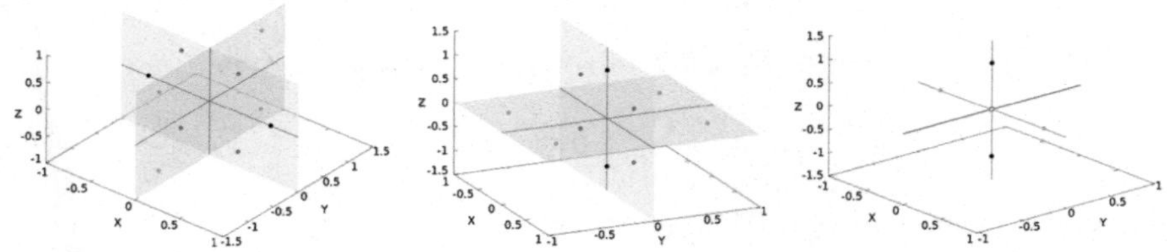

Fig. 4.2 Location of equilibria in normal case (left), radial case (middle) and tangential case (right)

In the tangential case the whole $Y-$ axis, except the origin, is of equilibrium points (3_T). In this case there are also four additional equilibria, two on the $Z-$ axis (1_T) and two on the $X-$ axis (2_T). In this chapter we will not study the degenerate equilibria with $H_T = 0\,(3_T)$.

4.2.6 Stability of the Equilibrium Points

We study the linear stability of the equilibrium points by looking at the eigenvalues of the Jacobian matrix of the differential equations at the equilibrium point. The results depend on the value of the angular quotient $\beta = n/\omega_c$. As a result, we numerically explore the evolution of the eigenvalues of the Jacobian as a function of β. We assume that the rotation rate of the dipole is such that only relative small values need to be considered, so we choose $\beta \in [-10, 10]$, except for the 1_N family, for which we consider $\beta \in [-25, 25]$.

Due to the symmetry of the system, the characteristic polynomial $p(\lambda)$ has only even-order terms, so it can be written as

$$p(\lambda) = \lambda^6 + b\lambda^4 + c\lambda^2 + d, \tag{4.15}$$

where b, c, and d are either functions of β or constants. The characteristic polynomials of the different equilibrium points are:

$$\begin{array}{lll}
1_N: & \lambda^6 + \left(\frac{\beta^2}{2} + 2\right)\lambda^4 + \dfrac{-\beta^2 + 8\beta + 5}{3}\lambda^2 + 6 & = 0,\\
2_N: & \lambda^6 + (14\beta^2 + 12\beta + 8)\lambda^4 + \dfrac{-40\beta^2 + 80\beta - 1}{3}\lambda^2 - 60 & = 0,\\
3_N: & \lambda^6 + (9\beta^2 + 12\beta + 8)\lambda^4 + (90\beta^2 + 120\beta - 47)\lambda^2 - 270 & = 0,\\
1_R: & \lambda^6 + \beta^2\lambda^4 - (6\beta^2 + 3)\lambda^2 + 18 & = 0,\\
2_R: & \lambda^6 + \left(\dfrac{9}{2}\beta^2 + 2\right)\lambda^4 + (9\beta^2 - 17)\lambda^2 - 18 & = 0,\\
3_R\,(Z = X): & \lambda^6 + (10\beta^2 - 12\beta)\lambda^4 + (24\beta^2 - 48\beta - 15)\lambda^2 + 36 & = 0,\\
3_R\,(Z = -X): & \lambda^6 + (10\beta^2 + 12\beta)\lambda^4 + (24\beta^2 + 48\beta - 15)\lambda^2 + 36 & = 0,\\
1_T: & \lambda^6 + \beta^2\lambda^4 + (3 - 3\beta^2)\lambda^2 - 36 & = 0,\\
2_T: & \lambda^6 + (9\beta^2 + 8)\lambda^4 + (81\beta^2 - 65)\lambda^2 - 324 & = 0.
\end{array} \tag{4.16}$$

Denoting $\kappa = \lambda^2$, Eq. (4.15) can be written as a cubic polynomial,

$$q(\kappa) = \kappa^3 + b\kappa^2 + c\kappa + d. \tag{4.17}$$

The discriminant of $q(\kappa) = 0$ is given by (see Ref. [11])

$$\Delta = b^2 c^2 - 4 b^3 d - 4 c^3 + 18 b c d - 27 d^2. \tag{4.18}$$

The roots of $q(\kappa) = 0$ can be classified according to the value of Δ. If $\Delta > 0$, Eq. (4.17) has three distinct real roots; If $\Delta = 0$, it has a real root with multiplicity three; If $\Delta < 0$ it has one real root and two complex conjugates ones. Now, the eigenvalues of the Jacobian that determine the stability of the equilibrium point come into pairs, i.e., for real ones we have $\pm\lambda$ with $\lambda \in \mathbb{R}$, while for complex ones we have the conjugated pair $\lambda, \bar{\lambda}$ with $\lambda \in \mathbb{C}$.

Let us briefly recall these classical results: if the real part of all the eigenvalues of the Jacobian are negative, then the equilibrium point is linearly stable; and if at least one eigenvalue has positive real part then the equilibrium point is unstable. In both cases the equilibrium point is called hyperbolic. If all eigenvalues are real, and at least one of them is positive and at least one is negative, then the equilibrium point which is unstable is called a saddle. Furthermore, if at least one eigenvalue of the Jacobian matrix is zero or has a zero real part, then the equilibrium is said to be non-hyperbolic and the stability cannot be determined using only the linear approximation. When there is one pair of pure-imaginary eigenvalues, the equilibrium point is said to have a centre component. The invariant stable, unstable and centre manifolds associated

Table 4.3 Dimensions of the stable W^s, unstable W^u, and centre W^c manifolds associated with all kinds of equilibria, as a function of β

Equilibrium label	β-range	dim $W^{s,u}$	dim W^c
$1_N, 1_R, 1_T$	$(-\infty, +\infty)$	2	2
$2_N, 3_N, 2_R, 2_T$	$(-\infty, +\infty)$	1	4
3_R $(Z = X)$	$(-\infty, -0.9516) \cup (3.4525, +\infty)$	0	6
	$(-0.9516, 3.4525)$	2	2
3_R $(Z = -X)$	$(-\infty, -3.4525) \cup (0.9516, +\infty)$	0	6
	$(-3.4525, 0.9516)$	2	2

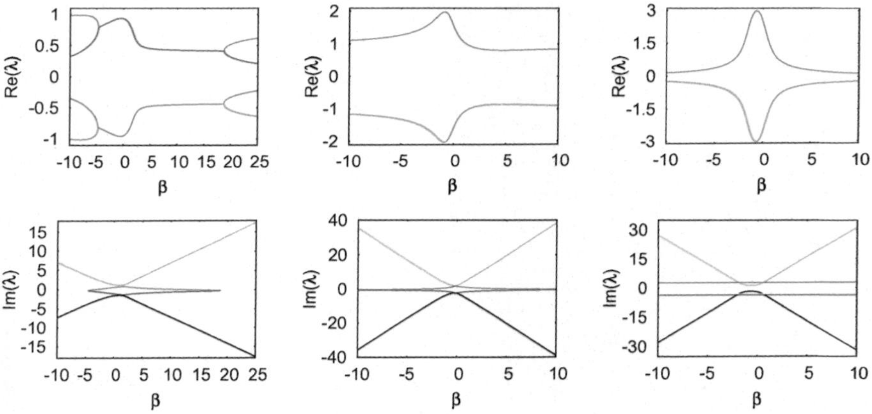

Fig. 4.3 Behaviour of the eigenvalues associated with the equilibrium points with labels 1_N(left), 2_N(middle) and 3_N(right) as a function of β. The top curves show the real part of the eigenvalues and the bottom ones show the imaginary part. Only the non-zero eigenvalues are shown

with the equilibrium are tangent to their linear approximations. Their dimension coincides with the number of eigenvalues with real part less than zero, greater than zero, and equal to zero, respectively. Table 4.3 lists the dimension of the invariant manifolds associated with all the equilibria as a function of β.

Figures 4.3 and 4.4 show the real and imaginary parts of the six eigenvalues of the equilibrium points as a function of the parameter β.

The four equilibria with label 1_N are always of saddle $\times$ saddle $\times$ centre type, which means that they are always unstable and there is a one-parameter family of periodic orbits around each of the four equilibrium points. The same happens to the four equilibrium points with label 1_R and 1_T. Note that the equilibria with index $2_N, 3_N, 2_R$ and 2_T are always of saddle $\times$ centre $\times$ centre type; the dimension of their associated stable and unstable manifolds is one, while the dimension of the centre manifold is four. For all these points, there are two families of periodic orbits associated with them. It is also observed that one pair of pure imaginary eigenvalues remain constant for 3_N, see Fig. 4.3 (right).

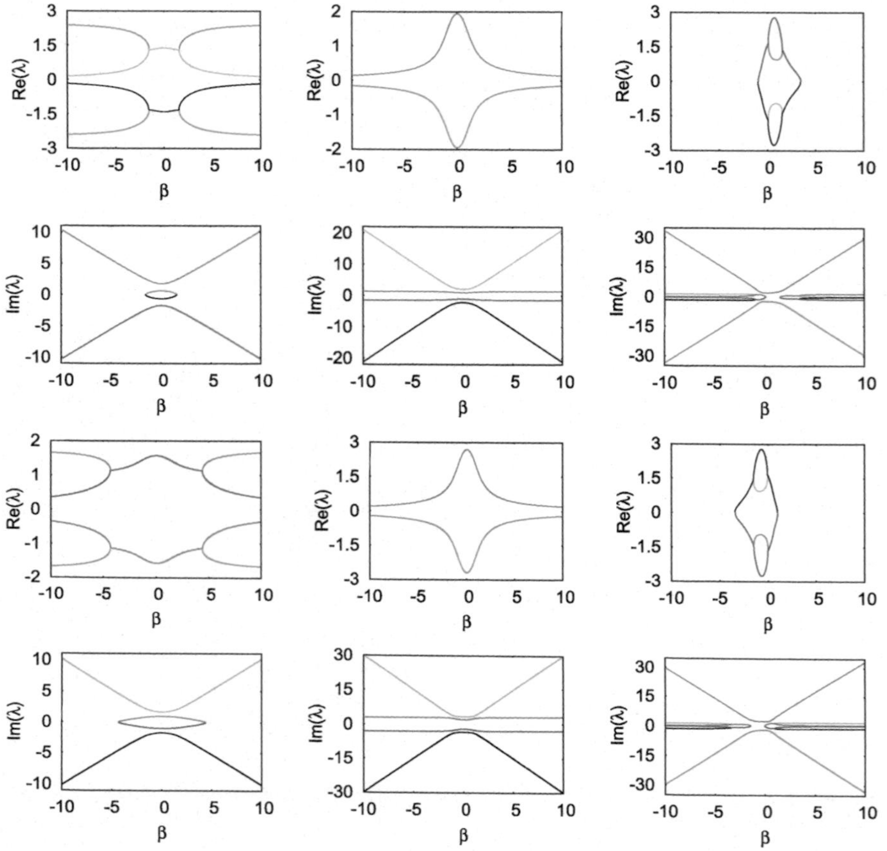

Fig. 4.4 The first two rows show the real part (first row) and the imaginary part (second row) of the eigenvalues associated to equilibria with labels 1_R(left), 2_R(middle) and 3_R with $Z = X$ (right) as a function of β. The last two rows display the behaviour of the eigenvalues associated to equilibria with labels 1_T(left), 2_T(middle) and 3_R with $Z = -X$ (right) as a function of β. In all the cases, only the non-zero eigenvalues are shown

As it is shown in Eq. (4.16), in the 3_R case we must consider different possibilities: if $Z = X$ and $\beta \in (-\infty, -0.9516) \cup (3.4525, \infty)$, or if $Z = -X$ and $\beta \in (-\infty, -3.4525) \cup (0.9516, \infty)$, all the equilibrium points are totally elliptic and, therefore, there exist three families of periodic orbits. For the other values of β, the 3_R points are of saddle $\times$ saddle $\times$ centre type, with only one family of periodic orbits. The existence of several families of periodic orbits enlarges the possibilities of being candidates as nominal trajectories in formation flight in future missions.

4.2.7 Zero Velocity Surfaces

In this section we analyse the zero velocity surfaces (ZVS) for all the three cases under consideration in order to better understand the structure of the phase and configuration spaces. The ZVS separate the phase space in two kinds of components: the forbidden region and the allowable region. We focus on energy levels around the equilibrium points, since the topology structure of ZVS qualitatively changes when the energy goes through these values.

According to Eqs. (4.9), (4.11) and (4.13), the general pattern of the first integrals of the three cases considered is

$$X'^2 + Y'^2 + Z'^2 = F(X, Y, Z) - H,$$

where the function $F(X, Y, Z)$ varies from one case to another, but the condition $F(X, Y, Z) - H \geq 0$ must be always fulfilled. The equation

$$F(X, Y, Z) - H = 0,$$

is the one that defines the ZVS.

To display the evolution of ZVS when the energy varies, for each kind of equilibrium point we plot the zero velocity surface at three close energy levels, i.e., one is the critical energy at the equilibrium point H_*, and the other two are $H_* - 0.05$ and $H_* + 0.05$. In order to see better the evolution of the ZVS with H, we show the ZVS and also one half of it (up to its intersection with a certain plane).

4.2.7.1 The Normal Case

Recall that, according to Table 4.2, in the normal case there are three kinds of equilibrium points: 1_N, 2_N and 3_N. We will denote their energy level by $H_{Ni}, i = 1, 2, 3,$ where i is the index associated with the label of the equilibria. If $\sigma > 0$ (positively charged follower) the number of the equilibria is eight, four of type 1_N and four of type 2_N. In Fig. 4.5 the behaviour of the ZVS is shown for the value of the energy H close to $H_{N1} = -1.587401$, which is the one associated to four equilibrium points: $(X, Y, Z) = (0, \pm\sqrt{2}Z, \pm(2/(3\sqrt{3}))^{1/3}) \approx (0, \pm 1.028721229, \pm 0.727415757)$.

In any of the regions of the configuration space (X, Y, Z) determined by the ZVS, the follower can only move inside the closed regions (such as the torus-shaped inner region surrounding the origin), as well as in the two unbounded regions that are below the lower and above the upper hyperbolic surfaces defined by the ZVS. As a result, there are three unconnected components of motion when $H_{N1} < -1.587401$. These three components meet at the equilibrium points when $H_{N1} = -1.587401$; when $H_{N1} > -1.587401$ there is only one connected component, where the motion can take place (the spacecraft is free to move in the whole configuration space connected through the 'bottle-necks' passages around the four equilibria).

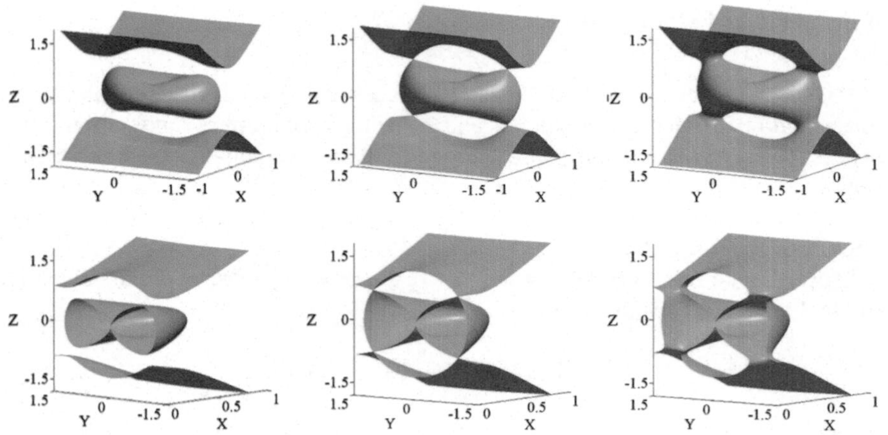

Fig. 4.5 For a positively charged follower ($q > 0$) in the normal case, evolution of ZVS as the energy level increases (from left to right, the values of H are $H_{N1} - 0.05$, $H_{N1} = -1.587401$, and $H_{N1} + 0.05$, respectively). In the three top figures the complete surfaces are represented, in the bottom ones they are displayed only up to their intersection with the $X = 0$ plane

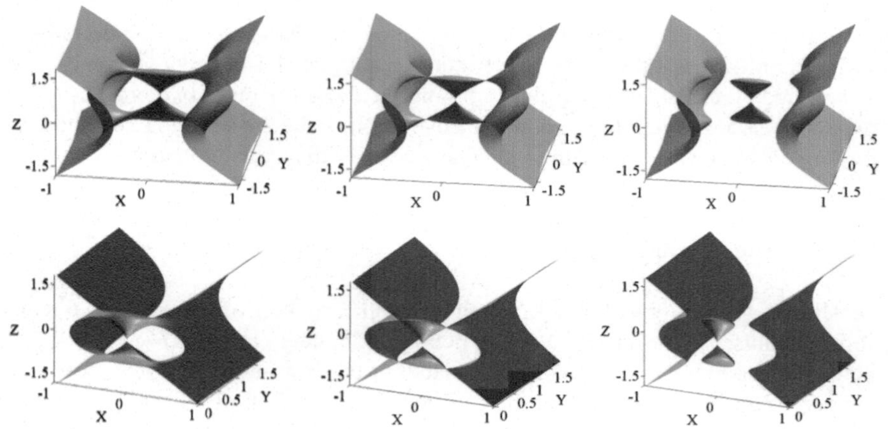

Fig. 4.6 For a positively charged follower ($q > 0$) in the normal case, evolution of ZVS as the energy level increases (from left to right, the values of H are $H_{N2} - 0.05$, $H_{N2} = -0.629960$, and $H_{N2} + 0.05$, respectively). In the three top figures the complete surfaces are represented and in the bottom ones they are displayed only up to their intersection with the $Y = 0$ plane

Figure 4.6 shows the evolution of the ZVS for the energy close to the one of the four equilibrium points with label 2_N, this is $H_{N2} = -0.629960$. In this case, when $H > H_{N2}$, the admissible region of motion has four components, two of them are closed and near the origin (above and below the $(X - Y)$ plane), and the other two are unbounded (to the right and to the left of the two hyperbolic surfaces). These four

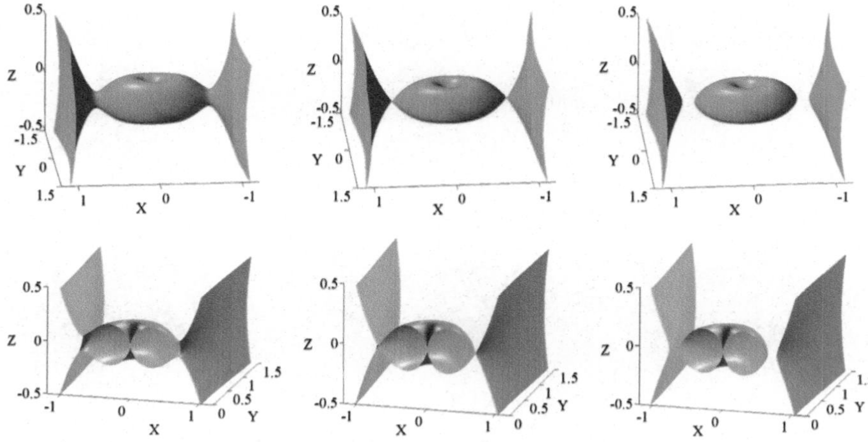

Fig. 4.7 For a negatively charged follower ($q < 0$) in the normal case, and from left to right, behaviour of the ZVS at the energy levels $H_{N3} - 0.05$, $H_{N3} = 4.326748$, and $H_{N3} + 0, 05$. The three bottom figures display the intersections of the ZVS with the $Y = 0$ plane

regions contact with each other at the equilibrium points when $H_{N2} = -0.629960$, and there is only one connected component of motion when $H < H_{N2}$.

If the follower is a negatively charged particle, the only kind of equilibrium is 3_N. The evolution of the ZVS for values of H close to $H_{N3} = 4.326748$ is the one shown in Fig. 4.7.

4.2.7.2 The Radial and Tangential Cases

In the radial case there are three kinds of equilibria. Two of the equilibrium points (both on the $Z-$axis) are of type 1_R, four equilibrium points on the $X - Y$ plane are of type 2_R, and four of type 3_R are on the $X - Z$ plane. The two equilibria with label 1_R are for positive charged particles ($q > 0$), while the four with both label 2_R and 3_R are for negatively charged spacecrafts ($q < 0$). The evolutions of the ZVS when the value of the energy H goes through the three critical values $H_{R1} = -3$, $H_{R2} = 2.289428$ and $H_{R3} = 1.88988$ are displayed in Fig. 4.8. The maximum number of connected components of the admissible regions of motion determined by the ZVS is three for the first two cases and four for the last one.

In the tangential case, as we already stated, we have only considered the equilibrium points of type 1_T and 2_T. According to Table 4.2, there are only two equilibrium points of each type with energy values equal to $H_{T1} = -3$ and $H_{T2} = 4.326748$. The evolutions of the ZVS for energy values close to these two are also displayed in Fig. 4.8.

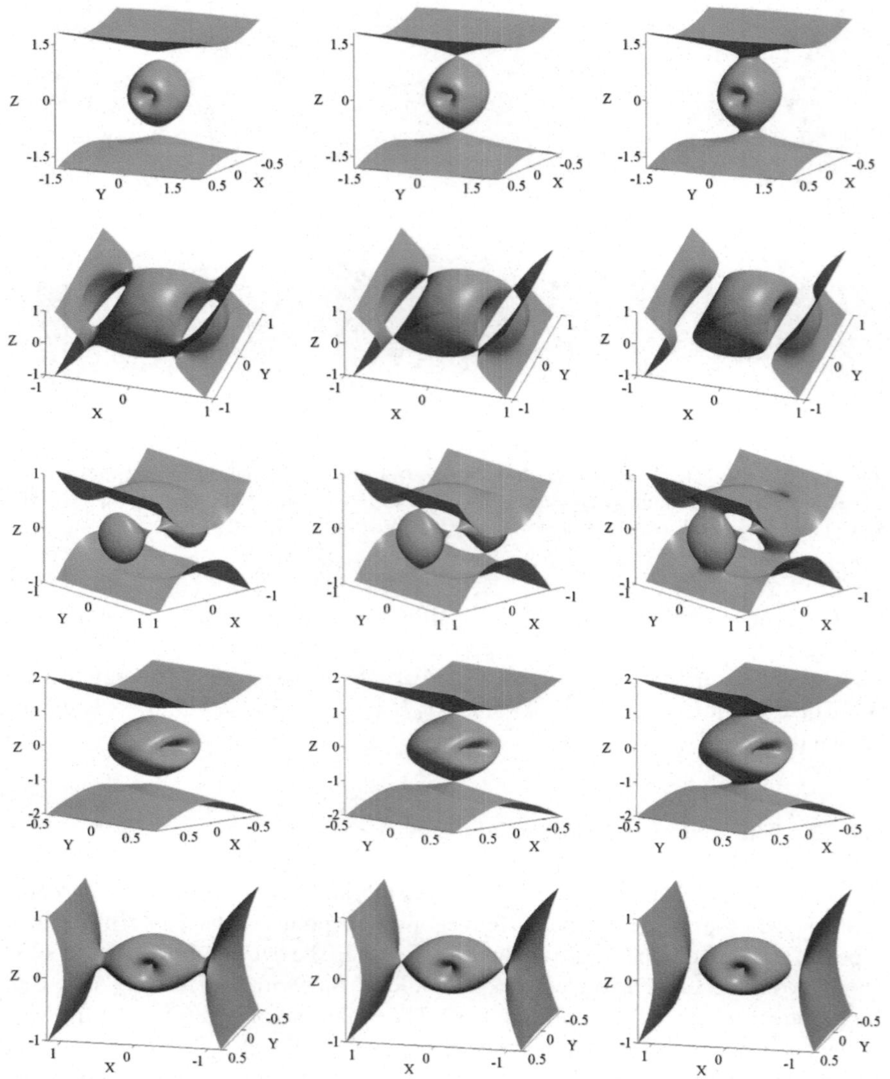

Fig. 4.8 From top to bottom, evolution of the ZVS near the energy values associated to the equilibrium points of type 1_R, 2_R, 3_R, 1_T and 2_T

4.3 Periodic and Quasi-periodic Orbits Emanating from Equilibria

According to the Lyapunov's centre theorem [18], the periodic orbits of the linearised equations of motion around an equilibrium point, which are associated with the pure imaginary eigenvalues λ_i of the Jacobian, persist when high order nonlinear terms are considered if the eigenvalues are incommensurable with each other. Furthermore, these orbits are embedded in a 1-parameter family of periodic orbits whose parameter can be chosen as the energy, the period, or any component of the state vector.

In this section we present the results of a systematic computation of families of symmetric and asymmetric periodic orbits of the systems defined by Eqs. (4.8), (4.10), and (4.12).

The computation of periodic orbits has been done using a numerical continuation method, based on a predictor followed by a differential corrector procedure. The initial guess used to start the procedure is given by the periodic solutions of the linearised equations. In most cases, the symmetries of the equations have been used to reduce the computational cost.

Around the equilibrium points we can also expect Lissajous orbits filling 2D invariant tori. We can imagine them as the coupling of two periodic motions with non-resonant frequencies ω_i and ω_j. Close to the equilibrium point, the frequencies of the tori will be close to the eigenvalues of the linearized system.However, the frequencies change with the amplitudes, and therefore they go across resonances. The proof of the existence of this Cantor set of tori follows the main idea underlying the KAM theorem (see [12]). We will compute them using a parameterisation method in Sect. 4.3.2.

4.3.1 Computation of Periodic Orbits Around the Equilibrium Points

Let $\lambda_1,\ldots,\lambda_6$ be the eigenvalues of any of the Jacobian matrices considered. The solution of the linearised system, around the associated equilibrium point, can be written as

$$\boldsymbol{x}(t) = c_1 e^{\lambda_1 t}\boldsymbol{v}_1 + c_2 e^{\lambda_2 t}\boldsymbol{v}_2 + \cdots c_n e^{\lambda_6 t}\boldsymbol{v}_6,$$

where $\boldsymbol{v}_i,\ i = 1,\ldots,6$ are the eigenvectors and $c_i,\ i = 1,\ldots,6$ are arbitrary constants.

We assume that the dimension of the centre manifold is, at least, two. Then, at least one eigenvalue λ_1 (together with its complex conjugate λ_4, since they appear in pairs) is pure imaginary and we can write $\lambda_{1,4} = \pm\sqrt{-1}s = \pm is$, with $s \in \mathbb{R}$. Setting $c_2 = c_3 = c_5 = c_6 = 0$ we get

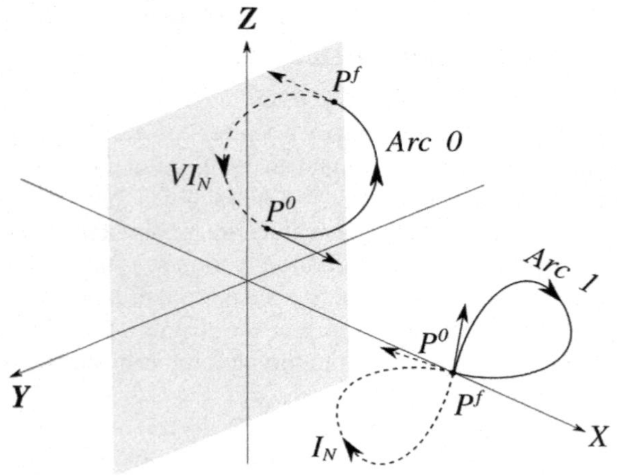

Fig. 4.9 Sketch of two kinds of symmetric periodic orbits: Arc 0 with coordinate plane symmetry, and Arc 1 with axial symmetry. The labels VI_N and I_N correspond to two of the symmetries defined in Table 4.1. These symmetries are used to save our computation because only half of the orbits are needed to be numerically integrated

$$\boldsymbol{x}(t) = c_1 e^{ist}\boldsymbol{v}_1 + c_4 e^{-ist}\bar{\boldsymbol{v}}_1,$$

where $\bar{\boldsymbol{v}}_1$ is the conjugate of $\boldsymbol{v}_1$.

Using this linear approximation, the periodic orbit used as initial guess for the non-linear system is

$$\boldsymbol{X}(t) = \boldsymbol{X}_{eq} + \epsilon \frac{\boldsymbol{x}(t)}{\|\boldsymbol{x}(t)\|},$$

where $\boldsymbol{X}_{eq}$ denotes the equilibrium point, ϵ is a small parameter, which usually takes the value of 10^{-3} in this work, and $\|\boldsymbol{x}(t)\|$ is the Euclidean norm of $\boldsymbol{x}(t)$.

Once the above initial orbits have been computed, we can start the predictor-corrector scheme for computing the family of periodic orbits associated to the eigenvalue λ_1, assuming that there are no other non-resonant imaginary eigenvalues ($\lambda_j/\lambda_1 \notin \mathbb{Z}$ for $j = 2, 3$). Of course, if the dimension of the centre manifold is larger than 2, the above procedure can be done for any other pair of imaginary eigenvalues.

We consider two main kinds of symmetric periodic orbits: those that are symmetric with respect to a coordinate plane, and those which have axial symmetry. Both kinds of orbits are sketched in Fig. 4.9. We consider only periodic orbits for which their initial and final configurations belong to the same kind of symmetry.

Let $(X^0, Y^0, Z^0, X'^0, Y'^0, Z'^0) \equiv (X_1^0, X_2^0, X_3^0, X_4^0, X_5^0, X_6^0)$ be the initial conditions of any of these symmetric orbits. According to the symmetry class (as given in Table 4.1) and the symmetry element (line or plane), if the initial condition is chosen at the symmetry element, only three components of the initial condition are different from zero. Moreover, the final point after one half period must be at the segment, with its velocity perpendicular to it.

If the symmetry element is the $Y - Z$, the $X - Z$ or the $X - Y$ plane, then $X = 0$, $Y = 0$ or $Z = 0$. In these cases, the initial velocity must be perpendicular to the

Table 4.4 Non-zero components i, j, k of the initial conditions for different kinds of symmetries considered

Case	Symmetry element	Non-zero components i, j, k
1	$X-$ axis	1, 5, 6
2	$Y-$ axis	2, 4, 6
3	$Z-$ axis	3, 4, 5
4	$Y-Z$ plane	2, 3, 4
5	$X-Z$ plane	1, 3, 5
6	$X-Y$ plane	1, 2, 6

symmetry planes, so $Y' = Z' = 0$, $X' = Z' = 0$ or $X' = Y' = 0$, respectively. If the symmetry element is the X, Y or Z axis, then the initial condition $Y = Z = 0$, $X = Z = 0$ or $X = Y = 0$, and also $X' = 0$, $Y' = 0$ or $Z' = 0$, respectively. For different kinds of symmetries, the non-zero components i, j, and k, of both the initial condition and the final condition after half a period, are given in Table 4.4.

4.3.1.1 The Differential Corrector

We use Case 4 of Table 4.4 as an example. For the other cases, the differential corrector method can be easily adapted. In this situation, the initial conditions are $(0,\ Y^0,\ Z^0,\ X'^0,\ 0,\ 0)^T$. We want that the orbit with these initial conditions intersects again the symmetry plane $X = 0$ with the same kind of behaviour. This means that after half a period (at $X = 0$) the following final conditions must be satisfied: $Y'^f = Z'^f = 0$.

Denoting by $\boldsymbol{X} = (X_i^0, X_j^0, X_k^0)^T$ the vector with the non-zero components of the initial condition, we want to solve the following system of non-linear equations:

$$\begin{cases} F_1(X_i^0, X_j^0, X_k^0) = X_m^f = 0, \\ F_2(X_i^0, X_j^0, X_k^0) = X_n^f = 0, \end{cases} \tag{4.19}$$

where $(X_1^f, X_2^f, X_3^f, X_4^f, X_5^f, X_6^f)$ denotes the final condition, (m, n) are the indexes associated to two of its components that must be zero. If the symmetry element is a plane, then $m = i + 3$ and $n = j + 3$. In case it is an axis, the values of the indices m and n depend on how the final condition is chosen after half a period. It can be either $X_{j-3}^f = 0$ or $X_{k-3}^f = 0$, so the periodicity condition to be satisfied is $X_{i+3}^f = 0$ (velocity orthogonal to the axis) and $X_{k-3}^f = 0$ or $X_{j-3}^f = 0$. Thus, if $X_l^f = 0$ is the final condition, then we have $l = k - 3$, $m = i + 3$, $n = j + 3$ for the plane symmetry case, and $l = j(k) - 3$, $\mathrm{m} = \mathrm{i} + 3$, $n = k(j) - 3$ for axial symmetry case.

The above system Eq. (4.19) of two equations with three unknowns can be written as

$$F(\boldsymbol{X}) = \boldsymbol{0}.$$

This system is solved by means of a modified Newton's method in which, at each iteration, we minimise the Euclidean norm of the correction given by Newton's method. For $K = 1, 2, 3, \ldots$ the equations of the iterative procedure are

$$\begin{aligned} \boldsymbol{X}_K &= \boldsymbol{X}_{K-1} + \Delta\boldsymbol{X}_{K-1}, \\ \Delta\boldsymbol{X}_{K-1} &= -G^T(GG^T)^{-1} \cdot F(\boldsymbol{X}_{K-1}), \end{aligned} \tag{4.20}$$

where G is the Jacobian matrix of F with respect to (X_i^0, X_j^0, X_k^0), and is given by

$$G = \begin{bmatrix} \Phi_{m,i} & \Phi_{m,j} & \Phi_{m,k} \\ \Phi_{n,i} & \Phi_{n,j} & \Phi_{n,k} \end{bmatrix} - \frac{1}{X_l'^f} \cdot \begin{bmatrix} X_m'^f \\ X_n'^f \end{bmatrix} \begin{bmatrix} \Phi_{l,i} & \Phi_{l,j} & \Phi_{l,k} \end{bmatrix},$$

where, Φ is the 6×6 state transition matrix and solution of the linear variational equations after half a period. The iterative procedure defined by Eq. (4.20) finishes when either $\|F(\boldsymbol{X}_K)\|$ or $\|\Delta\boldsymbol{X}_{K-1}\|$ is less than a certain threshold.

When none of the centre eigenvectors satisfy any of the symmetric configurations given in Table 4.4, the above refinement strategy is no longer valid. In this case the periodicity condition must be adjusted to return to the same initial state after a full period. In this case we have 7 unknowns $\boldsymbol{X} = [X^0, Y^0, Z^0, X'^0, Y'^0, Z'^0, T]^T$ (6 state vector + period), and 6 equations $\boldsymbol{X}^f - \boldsymbol{X}^0 = \boldsymbol{0}$.

To set the initial condition, we fix one of its components, for example: $Y^0 = 0$. Considering that the energy is preserved during the integration, we can eliminate one of the constraints on final state, i.e. $Z'^f - Z^0 = 0$, and as a consequence we have a system of 6 unknowns and 5 equations. To solve this non-linear system we proceed as in the symmetric case, requiring at each iteration for a minimum norm correction. Now, the Jacobian G in Eq. (4.20) is the 6×5 matrix $\Phi - Id$, of which we eliminate the 6th row, because of the constraint $Z'^f - Z^0 = 0$, and the second column, to remove the correction on initial component Y^0.

For the numerical integration and refinement we use a Runge–Kutta–Fehlberg 7–8 method with the following parameters: maximum local truncation error of the RKF78, 10^{-13}; maximum error in the determination of the final condition on the fixed plane/line, 10^{-11}; and maximum value of $\|F(\boldsymbol{X}_K)\|$ or $\|\Delta\boldsymbol{X}_{K-1}\|$, 10^{-10}.

4.3.1.2 The Continuation Method

The computation of one-parameter families of periodic orbits defined by their initial conditions $(X_i^0(s), X_j^0(s), X_k^0(s))$ as a function of a certain parameter requires a continuation method. We have chosen the arc-parameter, s. The curve $(X_i^0(s), X_j^0(s), X_k^0(s))$, in the 3-dimensional space (X_i, X_j, X_k), is the *characteristic curve* of the family and satisfies the system of differential equations (see [7]),

$$\frac{dX_i}{ds} = \frac{A_1}{A_0}, \quad \frac{dX_j}{ds} = \frac{A_2}{A_0}, \quad \frac{dX_k}{ds} = \frac{A_3}{A_0}, \tag{4.21}$$

where $A_0 = (A_1^2 + A_2^2 + A_3^2)^{1/2}$, $A_1 = (F^1_{X_j} F^2_{X_k} - F^1_{X_k} F^2_{X_j})$, $A_2 = -(F^1_{X_i} F^2_{X_k} - F^1_{X_k} F^2_{X_i})$, $A_3 = -(F^1_{X_i} F^2_{X_j} - F^1_{X_j} F^2_{X_i})$, and $F^{1(2)}_{X_i(X_j,X_k)}$ is the partial derivative of $F^{1(2)}$ with respect to $X_i(X_j, X_k)$.

The integration of Eq. (4.21) is done using an Adams–Bashforth method with one, two, three or four steps, depending on the number of available points on the curve. Therefore, supposing for instance, that four points $\boldsymbol{X}^0_r$, $\boldsymbol{X}^0_{r+1}$, $\boldsymbol{X}^0_{r+2}$, $\boldsymbol{X}^0_{r+3}$ are known, the integration of Eq. (4.21) will provide a new point $\boldsymbol{X}_{r+4}$ near the curve. Due to the relative low accuracy of this integration procedure, the resulting point $\boldsymbol{X}_{r+4}$ must be refined by means of the differential correction method previously explained in order to get the accurate initial conditions $\boldsymbol{X}^0_{r+4}$ of the new periodic orbit.

Considering the fixed step size Δs used in Adams–Bashforth method for the integration of Eq. (4.21), we have followed the automatic control strategy given in [28]. From one side, Δs must be adjusted to make sure that the curvature of the characteristic curve is not too strong. That is, the angle between the two segments determined by three consecutive points is less than some bound (we have used 0.1rad). From the other side, if the number of iterations required by Newton's method is small (less than 3 in our work), we can speed up the computation of the characteristic curve by increasing Δs. Analogously, if the number of Newton's iteration is too large (greater than 8) it is convenient to decrease Δs. The above step-size control must be implemented accurately, adding or removing points of the characteristic curve, since Adams–Bashforth method is a fixed step-size propagator.

We stop the continuation procedure if:

- The required step size Δs is smaller than a certain amount (10^{-6}), which usually happens when the family is close to a bifurcation or to its natural termination (for instance, at another equilibrium point).
- The size of the orbits becomes very large (modulus of the position vector larger than 50 units). This is because we are mainly interested in the application of the orbits obtained to formation flying, in which the follower is usually required to be in the proximity of the leader.

We remark that the arc-length continuation procedure is able to jump some bifurcations and remain along the original family. There are also possibilities that the procedure follows the bifurcated families, which must be analysed carefully by means of the stability parameters of the orbits.

4.3.1.3 Linear Stability of Periodic Orbits and Bifurcations

The linear stability of periodic orbits is given by the eigenvalues of the monodromy matrix M, this is, the variational matrix after one full period T. Recall that the eigenvalues of M, also called *characteristic multipliers*, are such that 1 is always a multiplier of at least multiplicity 2, and if λ is an eigenvalue, then $1/\lambda$, $\bar{\lambda}$ and $1/\bar{\lambda}$ are also eigenvalues with the same multiplicity (see [37]).

Since the systems under consideration have 6 degrees of freedom, we have three couples of multipliers that we denote as: $(\lambda_1, \lambda_1^{-1})$, $(\lambda_2, \lambda_2^{-1})$ and $(\lambda_3 = \lambda_3^{-1} = 1)$.

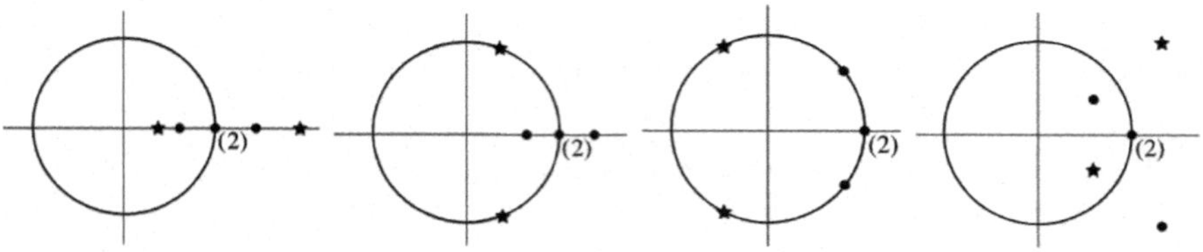

Fig. 4.10 Four possible configurations of the multipliers associated to the periodic orbits with respect to the unit circle in the complex plane. The stars correspond to the eigenvalues $(\lambda_1, \lambda_1^{-1})$ and the dots to $(\lambda_2, \lambda_2^{-1})$. The B_1 configuration corresponds to a real saddle, B_2 to a centre saddle, B_3 to a totally elliptic centre and B_4 to a complex saddle with a complex quadrupole of multipliers

The stability of the periodic orbits is given by the values of two traces: $Tr_i = \lambda_i + 1/\lambda_i, i = 1, 2$. If $Tr_1, Tr_2 \in \mathbb{R}$, then the periodic orbit is unstable when $|Tr_i| > 2$ for $i = 1$ or $i = 2$, and linearly stable if $|Tr_i| < 2$ for $i = 1, 2$. If $Tr_1, Tr_2 \in \mathbb{C}\backslash\{\mathbb{R}\}$ with $|Tr_i| \neq 1$, the orbit is complex unstable and the multipliers are a complex quadruple.

The four general configurations of the multiplies are displayed in the complex plane using Fig. 4.10. The B_3 configuration corresponds to a stable periodic orbit, with two couples of complex eigenvalues of modulus one, while the other three types correspond to unstable orbits. Transitions may happen between any two configurations, except $B_2 \leftrightarrow B_4$ and $B_1 \leftrightarrow B_3$. The cases where the multiplier 1 has multiplicity 4 or 6 are not shown since they are either degenerated cases, or transient cases, that correspond to a bifurcation or even a termination. One of these degenerated cases will appear in Sect. 4.3.5.2 and in one of the families of periodic orbits associated to the equilibrium points 2_R.

A bifurcation along a family of periodic orbits happens when there is a change of stability and this includes three cases (see [18]): *saddle-node bifurcation*, when one pair of real multipliers collide and move to the unit circle, this is, one trace crosses the value $+2$; *period doubling bifurcation*, when there are multipliers crossing the value -1, this is, one trace crosses the value -2; and *Krein bifurcation*, when two pairs of multipliers on unit circle collide and go to the complex plane ($B_3 \to B_4$), this is, the two traces with $|Tr_i \in \mathbb{R}| < 2$ for $i = 1, 2$ merge and become complex conjugate.

4.3.2 Computation of 2D Invariant Tori

In this section we explain the method used for the computation of 2D invariant tori around periodic orbits of elliptic type. This kind of orbits appear, for instance, around the periodic orbits associated to the equilibria with label 3_N (see Fig. 4.18). The numerical results obtained with this method will be shown later.

There are two commonly used parameterisation methods for the computation of 2D invariant tori [3]: the *stroboscopic map method* [6], that we use in our implementation, and the *Poincaré map method* [14], which, according to [9], is also called *large matrix methods*. We will show the connection between them.

Consider, for instance, Eq. (4.8) of the normal case, and write these equations as the first order system,

$$\dot{\chi} = f(\chi), \quad \chi = (X, Y, Z, X', Y', Z') \in \mathbb{R}^6. \tag{4.22}$$

According to [6], the computation of a 2D torus can be done by looking for a parameterisation of it, $\psi(\xi, \eta)$, with

$$\begin{array}{rccc} \psi : & \mathbb{R}^2 & \longrightarrow & \mathbb{R}^6 \\ & (\xi, \eta) & \longrightarrow & \psi(\xi, \eta), \end{array}$$

where ψ is 2π−periodic in both arguments ξ, η, and invariant under the flow associated to Eq. (4.22). This is:

$$\psi(\xi + t\omega_1, \eta + t\omega_2) = \phi_t\big(\psi(\xi, \eta)\big), \quad \forall t \in \mathbb{R}, \quad \forall \xi, \eta \in [0, 2\pi], \tag{4.23}$$

where ω_1, ω_2 are the frequencies of the torus, and ϕ_t denotes the flow associated to the differential system. So $\phi_t\big(\psi(\xi, \eta)\big)$ is the image under the flow of the point $\psi(\xi, \eta)$ after t time units.

In order to reduce the dimension of the problem, one of the parameters can be fixed, for instance taking $\eta = \eta_0 = 0$. This is equivalent to fix a curve on the torus, $\varphi(\xi) = \psi(\xi, 0)$, that must be invariant under $\phi_{2\pi/\omega_2}$. So we have,

$$\varphi(\xi + \rho) = \phi_T(\varphi(\xi)), \tag{4.24}$$

where $\rho = 2\pi\omega_1/\omega_2$ is the rotation number in radians of the invariant curve φ, and $T = 2\pi/\omega_2$. Once the parameterisation of the curve φ has been computed, the torus ψ can be determined by means of (see [6])

$$\psi(\xi, \eta) = \phi_{\frac{\eta-\eta_0}{2\pi}T}\left(\varphi\left(\xi - \frac{\eta - \eta_0}{2\pi}\rho\right)\right). \tag{4.25}$$

Let us assume that the invariant curve $\varphi : \mathbb{S}^1 \to \mathbb{R}^6$ that we want to compute has rotation number ρ, and its truncated Fourier expansion is,

$$\varphi(\xi) = \boldsymbol{C}_0 + \sum_{k=1}^{M} \big(\boldsymbol{C}_k \cos(k\xi) + \boldsymbol{S}_k \sin(k\xi)\big), \tag{4.26}$$

where $\boldsymbol{C}_0 \in \mathbb{R}^6$, $\boldsymbol{C}_k \in \mathbb{R}^6$, and $\boldsymbol{S}_k \in \mathbb{R}^6$ are Fourier coefficients to be determined.

The invariance equation for φ can be written as,

$$\varphi(\xi + \rho) = \phi_{T(\xi)}\big(\varphi(\xi)\big), \tag{4.27}$$

where the return-time $T(\xi)$ is fixed to be the period (or the normal period) of the periodic orbit for the *stroboscopic map method*. For the *Poincaré map method*, it depends on ξ and is different at each point.

4.3.2.1 Indeterminations of the Parameterisation

There are two indeterminations in the Fourier representation of $\varphi(\xi)$. One is due to the location of the invariant curve on the torus, since for any value of $\eta_0 \in [0, 2\pi]$, the curve $\varphi(\xi) = \psi(\xi, \eta_0)$ satisfies the invariance Eq. (4.24). This indetermination can be avoided by fixing one coordinate of $\boldsymbol{C}_0$. We note that this coordinate needs to be chosen taking into account the geometry of the torus.

The other indetermination is due to the possible phase shifts ξ_0 in the Fourier representation. This is, if $\varphi(\xi)$ satisfies the invariant equation (4.27), then $\varphi(\xi + \xi_0)$, $\forall \xi_0 \in \mathbb{R}$ also does. In fact we have,

$$\varphi(\xi + \xi_0) = \boldsymbol{C}_0 + \sum_{k=1}^{N_f} \left(\tilde{\boldsymbol{C}}_k \cos(k\xi) + \tilde{\boldsymbol{S}}_k \sin(k\xi) \right), \tag{4.28}$$

with the following representation, $\tilde{\boldsymbol{C}}_k = \boldsymbol{C}_k \cos(k\xi_0) + \boldsymbol{S}_k \sin(k\xi_0)$ and $\tilde{\boldsymbol{S}}_k = \boldsymbol{S}_k \cos(k\xi_0) - \boldsymbol{C}_k \sin(k\xi_0)$.

The phase shift indetermination can be avoided by setting one component of $\boldsymbol{C}_1$ or $\boldsymbol{S}_1$ equal zero with an appropriate selection of the value of ξ_0 (see [6]), and with the Fourier coefficients updated to $\tilde{\boldsymbol{C}}_k, \tilde{\boldsymbol{S}}_k, k = 1 \div N_f$, by means of Eq. (4.28). The index J of the component is chosen such that,

$$\|\boldsymbol{C}_1^J, \boldsymbol{S}_1^J\|_2 = \max_{j=1,2} \|\boldsymbol{C}_1^j, \boldsymbol{S}_1^j\|_2.$$

Between $\boldsymbol{C}_1^J$ and $\boldsymbol{S}_1^J$ we choose the one with maximum absolute value. Once both selections have been done, the unknown $\boldsymbol{C}_1^J$ (or $\boldsymbol{S}_1^J$) and the corresponding column can be removed from the Jacobian.

4.3.2.2 The System of Equations

The saddle component of a 2D torus, if it exists, introduces instability in the numerical integration. To overcome this problem, a multiple shooting procedure is used (see Ref. [6]). In this way, instead of computing only $\varphi(\xi)$, several curves, $\varphi_j(\xi)$, $j = 0 \div m - 1$, are computed. They are defined by,

$$\begin{aligned} \varphi_{j+1}(\xi) &= \phi_{T/m}\varphi_j(\xi), \quad j = 0 \div m - 2, \\ \varphi_0(\xi + \rho) &= \phi_{T/m}\varphi_{m-1}(\xi), \end{aligned} \tag{4.29}$$

where φ_0 is the original curve, and the value of m depends on the instability of the torus. We have used values of m between 2 and 4. All the curves are evaluated in a discrete set of points,

$$\xi_i = i\,\frac{2\pi}{1+2N_f}, \quad i = 0, \dots, 2N_f.$$

Since we fix the value of the energy H of the torus, one more equation is added to the above system. In this way, the final system to be solved using the modified Newton's method is,

$$\begin{aligned} H(\varphi_0(0)) - H &= 0, \\ \varphi_{j+1}(\xi_i) - \phi_{T/m}\varphi_j(\xi_i) &= 0, \quad j = 0 \div m-2, \\ \varphi_0(\xi_i+\rho) - \phi_{T/m}\varphi_{m-1}(\xi_i) &= 0, \end{aligned} \tag{4.30}$$

where, in principle, the unknowns are ρ, T, H and the Fourier coefficients $\boldsymbol{A}$ of dimension $6m \times (2N_f+1)$. The coefficients $\boldsymbol{A}$ can be expressed as $\boldsymbol{A} = (\boldsymbol{A}^0, \dots, \boldsymbol{A}^j, \dots, \boldsymbol{A}^{m-1})$ with $\boldsymbol{A}^j = (\boldsymbol{C}_0^j, \boldsymbol{C}_1^j, \dots, \boldsymbol{C}_{N_f}^j, \boldsymbol{S}_1^j, \dots, \boldsymbol{S}_{N_f}^j)$ for $j = 0 \div m-1$.

Since we look for torus embedded in two-parameter families that are specified by at least two parameters among (ρ, T, H), we always remove one unknown, usually ρ or H.

4.3.2.3 Initial Guess for the Curve φ

The computation of an initial guess for the curve φ is done following the ideas given in [6]. We consider a periodic orbit with centre part, this is, with at least one pair of the associated eigenvalues on the unit circle, and use the linear flow around the elliptic periodic orbit to compute the initial seed for the invariant curve.

Let M be the monodromy matrix associated to an initial point $\boldsymbol{\chi}_0$ of the periodic orbit with period T_p, that is, $M = D\phi_{T_p}(\boldsymbol{\chi}_0)$. The linear flow around $\boldsymbol{\chi}_0$ is given by,

$$L_{\phi_{T_p}}^{\boldsymbol{\chi}_0}(\boldsymbol{\chi}) = \boldsymbol{\chi}_0 + M(\boldsymbol{\chi} - \boldsymbol{\chi}_0). \tag{4.31}$$

Let $\cos\gamma + i\sin\gamma$ be an unitary eigenvalue of M and $\boldsymbol{v} = \boldsymbol{v}_r \pm \boldsymbol{v}_i$ the associated eigenvector. Then we have,

$$M(\boldsymbol{v}_r + i\boldsymbol{v}_i) = \big(\cos\gamma + i\sin\gamma\big)(\boldsymbol{v}_r + i\boldsymbol{v}_i) = \boldsymbol{v}_r\cos\gamma - \boldsymbol{v}_i\sin\gamma + i\big(\boldsymbol{v}_r\sin\gamma + \boldsymbol{v}_i\cos\gamma\big), \tag{4.32}$$

which can be rewritten as,

$$M[\boldsymbol{v}_r\ \ \boldsymbol{v}_i] = [\boldsymbol{v}_r\ \ \boldsymbol{v}_i]R(\gamma), \quad \text{with} \quad R(\gamma) = \begin{pmatrix} \cos\gamma & \sin\gamma \\ -\sin\gamma & \cos\gamma \end{pmatrix}.$$

Let us define a curve close to the periodic orbit by means of,

$$\varphi(\xi) = \chi_0 + \epsilon\big(v_r \cos\xi - v_i \sin\xi\big), \tag{4.33}$$

where ϵ is the distance from φ to the periodic orbit, and $\xi \in [0, 2\pi]$.

The curve φ is invariant under $L^{\chi_0}_{\phi_{T_p}}$, since

$$\begin{aligned} L^{\chi_0}_{\phi_{T_p}}(\varphi(\xi)) &= \chi_0 + M\epsilon\big(v_r \cos\xi + v_i \sin\xi\big) \\ &= \chi_0 + M[v_r \;\; v_i]\begin{pmatrix} \cos\xi & \sin\xi \\ -\sin\xi & \cos\xi \end{pmatrix}\begin{pmatrix} \epsilon \\ 0 \end{pmatrix} \\ &= \chi_0 + [v_r \;\; v_i]R(\gamma)R(\xi)\begin{pmatrix} \epsilon \\ 0 \end{pmatrix} = \chi_0 + [v_r \;\; v_i]R(\theta + \xi)\begin{pmatrix} \epsilon \\ 0 \end{pmatrix} \\ &= \varphi(\xi + \gamma), \end{aligned} \tag{4.34}$$

where γ is the rotation number in radians, and the two basic frequencies of the associated torus can be approximated by $(\gamma/T_p,\ 2\pi/T_p)$.

We remark that the above Eq. (4.34) is the linearised version of the invariance Eq. (4.24) with T taken as the period of the periodic orbit. Since it defines an invariant curve, with T and ρ close to those of the periodic orbit, the initial guess of φ can be given by,

$$\begin{aligned} &H = H(\chi_0), \quad T = T_p, \quad \rho = \gamma, \\ &C_0 = \chi_0, \quad C_1 = \epsilon v_r, \quad S_1 = -\epsilon v_i, \quad C_k = S_k = 0 \quad \text{for} \quad k > 2. \end{aligned} \tag{4.35}$$

When the multiple shooting is applied, the Fourier coefficients with $k = 0, 1$ of the intermediate curves φ_j are given by,

$$\varphi_j(\xi) = \phi_{j\frac{T_p}{m}}\chi_0 + \epsilon\big(v_r^j \cos\xi - v_i^j \sin\xi\big), \tag{4.36}$$

where $v^j_{(r,i)} = D\phi_{jT_p/m} v^j_{(r,i)}$.

There is another option for the initial guess of the curve $\tilde{\varphi}(\eta)$. It is the one associated to the centre part of the periodic orbit with T taken as the normal period of the centre part. Since this option has not been used in our computations we omit the details on how to implement it. Further information can be found in [6].

4.3.2.4 The Differential Corrector

System Eq. (4.30) is solved iteratively using the equations given in Eq. (4.20). The iterations stop either when $\|F\| < 10^{-10}$ or when $\|\Delta X_k\| < 10^{-10}$ is satisfied.

The differential of the first energy equation w.r.t. to the unknowns (in fact, only C_0^0) can be obtained easily by the chain rule. The Jacobian G of the remaining equations is given by,

$$G = \big[G_A \; G_\rho(G_H) \; G_T\big],$$

where the last two columns are the differential w.r.t. two unknowns among (ρ, T, H), and G_A is the differential associated to the Fourier coefficients $\boldsymbol{A}$. This is,

$$G_A = \begin{bmatrix} G_A^1 \\ G_A^2 \\ \vdots \\ G_A^{6(2N_f+1)} \end{bmatrix},$$

where the sub-matrices G_A^i, $i = 0, \ldots, 2N_f$ evaluated at point ξ_i are given by,

$$G_A^i = \left.\frac{D\,\boldsymbol{F}}{D\,\boldsymbol{A}}\right|_{\xi_i} = \begin{bmatrix} \dfrac{D\,\varphi_{j+1}(\xi_i)}{D\,\boldsymbol{A}} - \Phi\dfrac{D\,\varphi_j(\xi_i)}{D\,\boldsymbol{A}} \\ \hdashline \dfrac{D\,\varphi_0(\xi_i+\rho)}{D\,\boldsymbol{A}} - \Phi\dfrac{D\,\varphi_{m-1}(\xi_i)}{D\,\boldsymbol{A}} \end{bmatrix}, \quad j = 0 \div m-2, \tag{4.37}$$

where Φ is the 6×6 state transition matrix evaluated after a time-interval T/m. When $m = 1$ (no multiple shooting is applied), only the second row remains. Both $\frac{D\varphi_0(\xi_i+\rho)}{D\,\boldsymbol{A}}$ and $\frac{D\,\varphi_j(\xi_i)}{D\,\boldsymbol{A}}$ are sparse matrices, where the components associated to the corresponding Fourier coefficients $\boldsymbol{A}^j$ (the remaining ones are zeros) can be expressed with only $2N_f + 1$ diagonal entries B_k:

$$\begin{bmatrix} B_0 & & & \\ & B_1 & & \\ & & \ddots & \\ & & & B_{2N_f} \end{bmatrix}.$$

The matrices B_i, of dimension $6 \times (2N_f + 1)$, can be expressed as,

$$B_i = \begin{bmatrix} B_{i1} & & \\ & B_{i2} & \\ & & \ddots \\ & & & B_{i6} \end{bmatrix}$$

where $B_{il} = [1 \;\; \cos(\alpha) \;\; \ldots \;\; \cos(N_f\alpha) \;\; \sin(\alpha) \;\; \ldots \;\; \sin(N_f\alpha)]$ for $l = 1 \div 6$; $\alpha = \xi_i + \rho$ and $\alpha = \xi_i$ for $\frac{D\,\varphi(\xi_i+\rho)}{D\,\boldsymbol{A}}$ and $\frac{D\,\varphi(\xi_i)}{D\,\boldsymbol{A}}$, respectively.

We note that during the iterative procedure the number of frequencies N_f of the truncated Fourier expansion can be changed according to the accuracy to be achieved. The value is chosen in such a way that the maximum norm of the last $N_f/4$ coefficients is one order of magnitude smaller than the required tolerance. If this condition is not fulfilled, we double it to $2N_f$ and initially setting $\boldsymbol{C}_k = 0$ and $\boldsymbol{S}_k = 0$ for $k = N_f + 1, \ldots, 2N_f$. We have usually started with $N_f = 16$, which is enough for a curve with a regular shape such as a circle or an ellipse. This value needs to be increased for curves with more complex shape. However, to avoid too large systems, we have set the maximum of N_f to 128.

As an important remark about the implementation of the described procedure we want to mention that the computation of the $\cos(k\alpha)$ and $\sin(k\alpha)$ values, for $k = 1, .., 2N_f$, has been done using the stable trigonometric recurrences given in [30]. This avoids low accuracies when the arguments are large, moreover it saves CPU time.

Another remark is that we could "fall back" to the starting periodic orbit, since the periodic orbit itself is also a solution of the system with $\boldsymbol{C}_0 = 0$, $\boldsymbol{C}_K = 0$, $\boldsymbol{S}_k = 0$. To overcome this problem, we have fixed a non-zero component of $\boldsymbol{C}_0$. This, in principle, only needs to be done for the computation of the first torus, since during the continuation, the rotation number ρ and the return time T are varied, and therefore they are different from the ones of the periodic orbit.

Finally, as it has already been previously stated, two more unknowns need to be removed to avoid the curve indetermination and the phase shift indetermination. Again, this can be done by eliminating the corresponding columns from G.

4.3.2.5 Continuation of the 2D Tori Family

We use a similar predictor-corrector scheme for the continuation of the tori family as the one described in Sect. 4.3.1.2 for the continuation of periodic orbits. Once an invariant 2D torus is computed close to an elliptic periodic orbit, the continuation is done along its tangential space, that is the kernel of the system $\text{Ker}(DF)$. Thus, let us assume a torus $\boldsymbol{X}^k$ is computed. The prediction step of a new torus is given by $\boldsymbol{X}^{k+1} = \boldsymbol{X}^k + \Delta s \cdot \boldsymbol{v}$, with $\boldsymbol{v} \subset \text{Ker}(DF)$, $\|\boldsymbol{v}\| = 1$, and then it is refined using the modified Newton's method.

We can embed the torus in an isoenergetic family by keeping fixed the energy H, and varying the value of ρ and T, or vice versa, keeping fixed the rotation number ρ and letting T and H vary. However we must note that the tori are embedded in a Cantorian family with gaps due to resonances. These gaps can be jumped through if we tune the continuation step carefully. However, in many cases, the continuation procedure may stop when close to a resonance where the shape of the curves becomes complicated.

We remark that, due to the phase shift indetermination, the system of equation used for prediction and correction are different (see [6]). For the differential correction of the first torus, three unknowns are removed, which include one coordinate C_0^I to avoid the curve indetermination, one coordinate C_1^J (or S_1^J) to avoid the phase shift indetermination, and one non-zero coordinate of $\boldsymbol{C}_0^K$ to avoid falling back to the staring periodic orbit. So the system is of dimension $(6m(2N_f + 1) + 1) \times \big(6m(2N_f + 1) - 1\big)$ and full-rank. It is solved using a general QR factorization.

For the prediction of a new torus, we remove only C_1^J to select a curve on the torus and use the remaining equations to compute $\text{Ker}(DF)$. The system is rank-deficient and of dimension $(6m(2N_f + 1) + 1) \times \big(6m(2N_f + 1) + 1\big)$. The computations of $\text{Ker}(DF)$ are carried out using QR factorization with column pivoting (LAPACK routine DGEQP3 [1]). Since the set of solutions is two-dimensional, among which we choose the one that is orthogonal to the direction of the phase shift indetermination, which is obtained from the differential of Eq. (4.28) w.r.t. ξ_0.

For the continued curves, it is not necessary to consider the falling back problem to the original periodic orbit, so we only need to remove two unknowns C_0^I and C_1^J. The system is also rank-deficient, and a QR with column pivoting based Linear Least Squares solver (LAPACK routine DGELSY [1]) is used to solve the system defined by Eq. (4.30) with least-norm corrections.

The value of the continuation step Δs is adjusted within a certain interval (for instance $[10^{-5}, 10^{-2}]$). If the number of is small (less than three), we multiply it by two. If the refinement fails to converge or the number of iteration is large (greater than eight), we divide it by two and start again the procedure using as prediction the tangential space of the last successfully computed torus. Moreover, to guarantee a smooth continuation, we require the angle between the last three curves to be less than a given tolerance (we use 15°). If it exceeds this value, we divide Δs by 2 and restart the prediction from the previous torus. For the computation of the angle, we consider two unknowns among (ρ, T, H) and the remaining constant term $\boldsymbol{C}_0$ despite of $\boldsymbol{C}_0^I$.

We repeat this process until one of the following conditions is satisfied:

1. The continuation step is smaller than the lower bound allowed (10^{-6}).
2. The magnitude of the Fourier coefficients with $k \geq 1$ is smaller than the accuracy required by the refinement (10^{-10}). This means we end up at a periodic orbit.
3. The number, N_f, of Fourier modes required is larger than the maximum value allowed (set as 128 in this work). This usually happens when we are close to the resonance or the shape of the curve becomes very complex. We are able to jump the resonance many times.

A trick to jump through possible resonances during the continuation is by tuning the step Δs and the value of N_f which can be applied when the procedure is convergent but N_f needs to be increased. During three iterations we simply allow to start again with N_f unchanged and the step size increased to $1.3\Delta s$. If the iterations fail to converge, we decrease $\Delta s = \Delta s/2$ and restart with $N_f = 2N_f$. Note that the value of N_f might change during the continuation, so the dimension of the Fourier coefficients $(\boldsymbol{C}_k, \boldsymbol{S}_k)$ needs to be specified carefully for the new curve.

4.3.2.6 The Relation Between the Two Parameterisation Methods for the Computation of 2D Invariant Tori

In this section we study the relation between the stroboscopic map method and the Poincaré map method for the computation of 2D invariant tori. The study is done by checking the return times required by both methods and assuming that no multiple shooting is used.

Let φ be an invariant curve computed using the stroboscopic map method and $\overline{\varphi}$ another one with the same rotation number associated to an elliptic periodic orbit.

We assume that they are close, so for each point $\varphi(\xi)$ on the curve φ there is a point $\overline{\varphi}(\xi)$ close to $\varphi(\xi)$ and on the curve $\overline{\varphi}$ such that,

$$\varphi(\xi) = \phi_{\tau(\xi)}\big(\overline{\varphi}(\xi)\big), \tag{4.38}$$

where $\tau(\xi)$ is a small time deviation (positive or negative) depending on ξ.

Looking at the left-hand side of the invariance equation Eq. (4.24), and using equation Eq. (4.38), we obtain,

$$\varphi(\xi+\rho) = \phi_{\tau(\xi+\rho)}\big(\overline{\varphi}(\xi+\rho)\big) = \phi_{\tau(\xi+\rho)} \circ \phi_{\overline{T}(\xi)}\big(\overline{\varphi}(\xi)\big) = \phi_{\tau(\xi+\rho)+\overline{T}(\xi)}\big(\overline{\varphi}(\xi)\big). \tag{4.39}$$

Similarly, substituting Eq. (4.38) in the right-hand side of Eq. (4.24) we get,

$$\phi_T\big(\varphi(\xi)\big) = \phi_T \circ \phi_{\tau(\xi)}\big(\overline{\varphi}(\xi)\big) = \phi_{T+\tau(\xi)}\big(\overline{\varphi}(\xi)\big). \tag{4.40}$$

Hence, the relation between T and $\overline{T}(\xi)$ is $\tau(\xi+\rho) + \overline{T}(\xi) = T + \tau(\xi)$, which can be rewritten as the homological equation,

$$\tau(\xi) - \tau(\xi+\rho) = \overline{T}(\xi) - T, \tag{4.41}$$

and since both sides of the above equation have zero-average, we have

$$T = \frac{1}{2\pi}\int_0^{2\pi} \overline{T}(\xi)d\xi. \tag{4.42}$$

The homological equation can be solved using Fourier series

$$\begin{aligned} \overline{T}(\xi) &= \sum_{k\in\mathbb{Z}} \overline{T}_k e^{2\pi i k\xi}, \quad & \tau(\xi) &= \sum_{k\in\mathbb{Z}} \tau_k e^{2\pi i k\xi}, \\ \overline{T}_0 &= T, & \tau_0 &= 0. \end{aligned} \tag{4.43}$$

where the $\overline{T}_0$ and τ_0 are the constant terms of the series. The coefficients $\overline{T}_k$ can be computed by evaluating $\overline{T}(\xi)$ in a regular grid and using a FFT. Moreover we have,

$$\begin{aligned} \tau(\xi) - \tau(\xi+\rho) &= \sum_{k\in\mathbb{Z}} \left(\tau_k e^{2\pi i k\xi} - \tau_k e^{2\pi i k(\xi+\rho)}\right) \\ &= \sum_{k\in\mathbb{Z}} \tau_k (1 - e^{2\pi i k\rho}) e^{2\pi i k\xi} \\ &= \sum_{k\in\mathbb{Z}} \overline{T}_k e^{2\pi i k\xi}. \end{aligned} \tag{4.44}$$

Hence for $k \neq 0$ we have,

$$\tau_k = \frac{\overline{T}_k}{1 - e^{2\pi i k\rho}}. \tag{4.45}$$

Since τ_0 is a free parameter, we can take $\tau_0 = 0$. If $(C_k^\tau, S_k^\tau) \in \mathbb{R}$ and $(C_k^T, S_k^T) \in \mathbb{R}$ denote the Fourier real coefficients of τ and $\overline{T}$, respectively, from Eq. (4.45)it follows that for $k \neq 0$,

$$C_k^\tau = \frac{1}{2}\left(C_k^T + S_k^T \cot(\tfrac{k\rho}{2})\right), \quad S_k^\tau = \frac{1}{2}\left(S_k^T - C_k^T \cot(\tfrac{k\rho}{2})\right). \tag{4.46}$$

As a final conclusion we can say that the two methods are closely related and one can easily go from one to the other through the return time. For the stroboscopic map method there are no geometric constraints in the selection of the invariant curve of the torus, which, on the other hand, introduces one more phase shift indetermination to deal with [6]. In the *Poincaré map method*, the initial curve is specified by the intersection of the torus with a prescribed surface of section, this section needs to be carefully chosen since the vector field of the torus must be transverse to it. From the computational point of view, we can say that the Poincaré map method requires at each step the solution of $4(2N_f + 1) \times \big(4(2N_f + 1) - 1\big)$ linear systems via the Newton's iteration method, while in the stroboscopic map method the dimension is $6(2N_f + 1) \times \big(6(2N_f + 1) - 1\big)$.

We prefer the stroboscopic map method over the Poincaré map method because the reliance on geometry makes it difficult to choose a suitable Poincaré section guaranteeing the transversality of the tori with the section, especially during the continuation.

4.3.3 Numerical Results on Periodic and Quasi-periodic Orbits

In this section we show the numerical results obtained in the computation of the families of periodic, and quasi-periodic orbits (tori) around the equilibrium points with an associated non-empty centre manifold. As in the preceding section, we discuss separately the three cases considered: normal, radial, and tangential.

4.3.4 The Normal Case

We recall that in the normal case, and for any value of the parameter β, we have three kinds of equilibria: 1_N, 2_N and 3_N. All the points of the first kind are of saddle $\times$ saddle $\times$ centre type, which means that there is one one-parameter family of periodic orbits around each of them. The other two kinds of equilibria, 2_N and 3_N, are saddle $\times$ centre $\times$ centre type and they have associated two families of periodic orbits.

The symmetries of the differential Eq. (4.8), previously discussed in Sect. 4.2.3, have been used for the computation of different families of symmetric periodic orbits.

4.3.4.1 Periodic Orbits Around the 1_N Equilibria

There are four equilibrium points of type 1_N located at $(0, \pm\sqrt{2}(2/(3\sqrt{3}))^{\frac{1}{3}}, \pm(2/(3\sqrt{3}))^{\frac{1}{3}})$. We will focus on the two with $Z > 0$, which are symmetric w.r.t the $Y = 0$ plane. Since both points are in the $Y - Z$ plane, for the computation of symmetric periodic orbits around them we have used the symmetry about this plane, labelled as VI_N in Table 4.1. We notice that only one of these families has to be computed, and the ones around the other equilibrium points are just symmetric images with respect to the $Z = 0$ plane.

For the computations that follow we have used $\beta = 2$ as reference value. For this value of β, the associated eigenvalues to the equilibrium points are $(\pm 0.6509 \pm 1.0291i, \ \pm 1.6520i)$. If $\boldsymbol{\omega}^c = \omega_r^c \pm \omega_i^c$ denotes the eigenvector associated to the purely imaginary eigenvalue, $\pm 1.6520i$, we have

$$\omega_r^c = [0.2477,\ 0,\ 0,\ -0.1968,\ 0.7250]^T, \quad \omega_i^c = [0,\ 0.1191,\ -0.4389,\ 0.4092,\ 0,\ 0]^T.$$

This is the data that has been used to determine the first orbit (in the linear approximation of the flow) for the computation of the family of periodic orbits, as explained in Sect. 4.3.1. The resulting refined periodic orbit around the equilibrium point with $Y > 0$ and $Z > 0$ is shown in Fig. 4.11.

Using the continuation procedure, and staring at a certain periodic orbit around the equilibrium point, we can compute the whole family of periodic orbits, as is shown in Fig. 4.12. The family starts at the equilibrium point with $Y > 0$ and ends at the symmetric point $(0, -\sqrt{2}(2/(3\sqrt{3}))^{\frac{1}{3}}, (2/(3\sqrt{3}))^{\frac{1}{3}})$ or, in other words, the families emanating from both points are connected at any intermediate orbit of the family, for instance, the one represented in black in the figure.

Fig. 4.12 also shows the characteristic curve (energy H versus period T) and the results of the stability analysis. Different colors denote different configurations of the multipliers of the periodic orbits. Two period-doubling bifurcations are detected along the family when the trace of the monodromy matrix crosses the value -2. They are indicated by two vertical lines.

The family starts at the right point of the characteristic curve, with $H_n = -1.587402$, which corresponds to the orbit shown in Fig. 4.11. The period of this orbit is very close to $\lambda = \pm 1.6520i$ which, according to Lyapunov theorem, is the limit value of the periods of the family. The first orbits of the family are of type B_4. The family transits to type B_1 when the energy goes below $H_n = -1.597089$. A period-doubling bifurcation occurs at $H_n = -1.614741$, after which the family

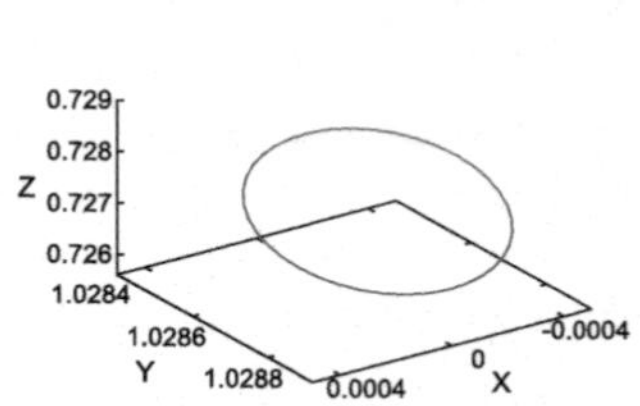

Fig. 4.11 For $\beta = 2$, periodic orbit near the equilibrium point $(0, \sqrt{2}(2/(3\sqrt{3}))^{\frac{1}{3}}, (2/(3\sqrt{3}))^{\frac{1}{3}})$ associated to $\lambda = \pm 1.6520i$

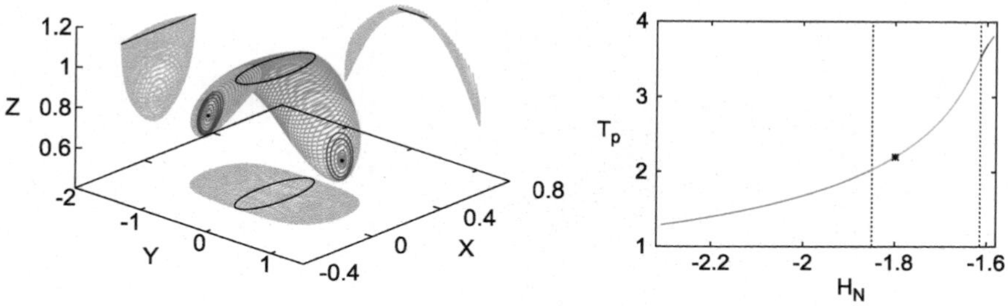

Fig. 4.12 For $\beta = 2$, 3D representation and 2D coordinate projections of the periodic orbits of the two connected families around the 1_N equilibrium points $(0, \pm\sqrt{2}(2/(3\sqrt{3}))^{\frac{1}{3}}, (2/(3\sqrt{3}))^{\frac{1}{3}})$ (left). The right had side plot displays the energy H versus period T characteristic curve of the family starting at the equilibrium point and finishing at the periodic orbit represented in black. The colors in both figures indicate different configurations of the multipliers of the periodic orbits: B_4 in red, B_1 in blue, B_2 in orange and B_3 in green. Only green orbits are stable and the vertical lines indicate the bifurcation points

goes to type B_1 as a real saddle. The family is unstable up to $H_n = -1.854221$, where a second period-doubling bifurcation is detected, and afterwards it is of type B_3 (green) and becomes stable. The black point in the characteristic curve indicates the periodic orbit from which we compute an iso-energetic family of 2D tori.

Due to the type VII_N symmetry, the orbits to the left of the middle (black) orbit –left branch– are the images of the orbits to the right of the middle orbit –right branch–. As a consequence, the characteristic curves of both branches are the same. The periodic orbit connecting the two branches (displayed in black) corresponds to the most left point of the characteristic curve, and the associated multipliers correspond to a degenerated centre with four multipliers equal to $+1$. Moreover, the middle (black) periodic orbit is symmetric with respect to both the $X - Z$ plane and the $Y - Z$ plane, and is "almost" parallel to the $X - Y$ plane; its maximum displacement along the Z direction is of the order of 1×10^{-5} non-dimensional units. These "almost" parallel orbits could be potential nominal orbits for a displaced observation mission.

The behaviour of these two families, as well as the described connection, is similar for all the positive values of β that have been explored, this is: when the dipole rotates in the same sense as the chief's motion around the Earth. When $\beta < 0$ the behaviour of the families changes.

For $\beta = -2$, Fig. 4.13 shows the behaviour of the orbits of the family associated to the equilibrium point with $Y > 0$ together with its characteristic curve. The continuation of this family has been stopped when $|Z| > 45$. It can be seen that, as the energy decreases, the orbits become very large along the Z direction and their period tends to a value close to π (we have not found an explanation of this fact). Now, the connection of both families disappeared, and one family is just the symmetric of the other without common orbits. According to the behaviour of the characteristic multipliers the orbits are always unstable, although their type follows the transitions: $B_4 \to B_1 \to B_4 \to B_1$. All families, for $\beta = 2$ and for $\beta = -2$, are shown in Fig. 4.14.

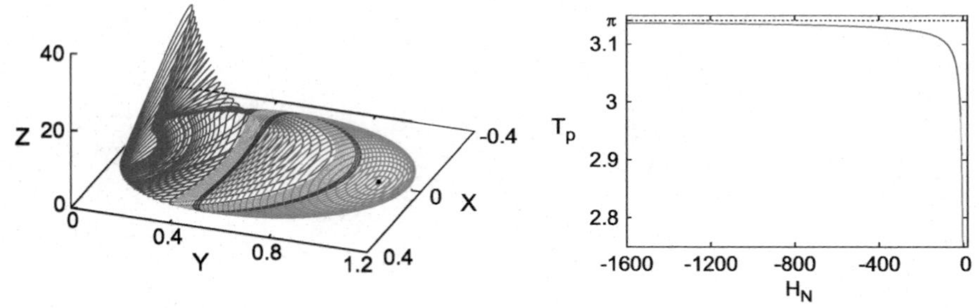

Fig. 4.13 For $\beta = -2$, 3D representation of the periodic orbits of the family around the 1_N equilibrium point $(0, \sqrt{2}(2/(3\sqrt{3}))^{\frac{1}{3}}, (2/(3\sqrt{3}))^{\frac{1}{3}}))$ (left). The right had side figure displays the energy H versus period T characteristic curve of the family starting at the equilibrium point. The colors in both figures indicate different configurations of the multipliers. Their types go through the following transitions: B_4 (red) to B_1 (blue), back to B_4 (orange) and B_1 (purple)

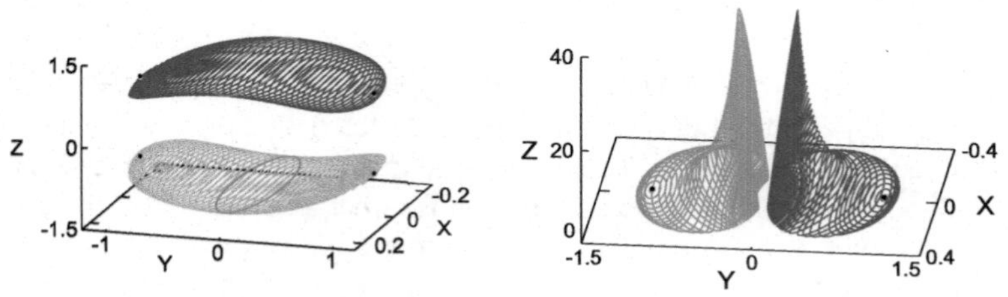

Fig. 4.14 3D representation of the families of periodic orbits around the equilibrium points of type 1_N. The left hand side plot shows the four families corresponding to $\beta = 2$. The right hand side shows the two families with $\beta = -2$ and $Z > 0$ (the other two families are symmetric w.r.t the $Z = 0$ plane)

4.3.4.2 Periodic Orbits Around the 2_N Equilibria

There are four equilibrium points of type 2_N with coordinates $(\pm(1/(16\sqrt{6})^{1/3}, 0, \pm\sqrt{5}(1/(16\sqrt{6})^{1/3})$. As in the preceding case, we mainly consider the point with $X > 0$ and $Z > 0$. The periodic orbits around the remaining points can be obtained using suitable symmetries. As it is shown in Sect. 4.2.4, the centre manifold associated to all these equilibria has dimension four and so it contains two families of periodic orbits. Figure 4.15 shows the orbits of both families together with their associated characteristic curves for the value $\beta = 2$. The VII_N symmetry w.r.t. the $X - Z$ plane has been used for their computation.

Analogous to what happens in the 1_N case, when $\beta > 0$ one of the families (top line of Fig. 4.15) has its termination at the equilibrium point symmetric with respect to the $X = 0$ plane and, in fact, the families around $(+(1/(16\sqrt{6})^{1/3}, 0, \pm\sqrt{5}(1/(16\sqrt{6})^{1/3})$ and $(-(1/(16\sqrt{6})^{1/3}, 0, \pm\sqrt{5}(1/(16\sqrt{6})^{1/3})$ are the same one. All the periodic orbits around these two points are symmetric w.r.t. the $X - Z$ plane, and the orbits around one point are the symmetric images of the ones around the other point w.r.t the $X = 0$

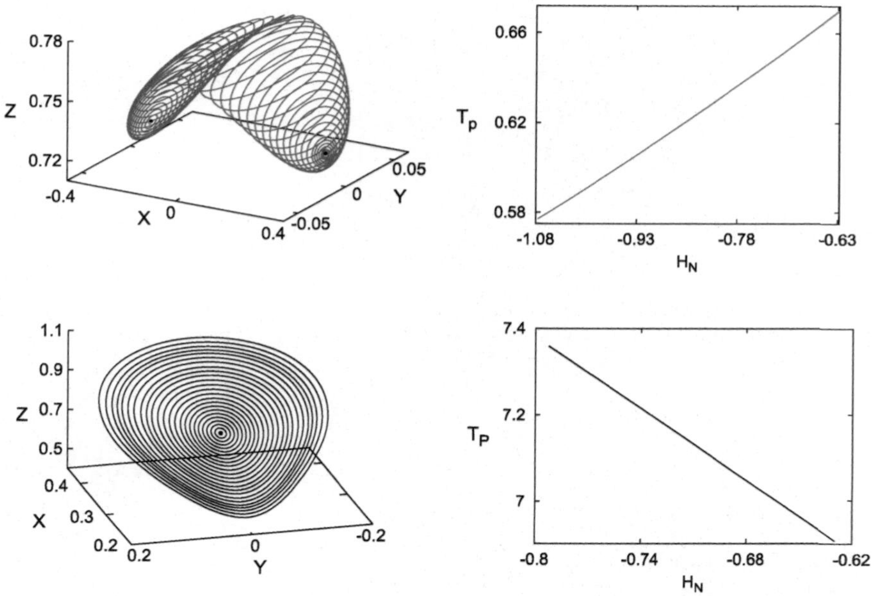

Fig. 4.15 For $\beta = 2$, the two families of periodic orbits emanating from the equilibrium point of type 2_N with $X > 0$ and $Z > 0$ (left), with the 3D representation of the orbits in the left hand side figure and the associated characteristic curves at the right hand side one. Both families are unstable (B_2)

plane (VI_N symmetry). As a consequence, the characteristic curves are just the one displayed in the figure but travelled twice. We note that the left point of this curve corresponds to the orbit connecting the two families and its multipliers are the ones of a degenerated centre. For the other family, in the bottom line of the figure, the left point of the characteristic curve corresponds to its termination, where the orbit degenerates into a saddle.

The characteristic curves of both families display a monotonous linear relation between the energy and the period, and all the orbits are unstable of type B_2 (saddle $\times$ centre).

Along the family, in the bottom line of Fig. 4.15, we have detected six saddle-node bifurcations. The orbits of the bifurcated families, and the one from which they are born, together with the associated characteristic curves are shown in Fig. 4.16, where each family is represented with a different color. All these families are unstable orbits, with multiplier configuration of type B_1 (saddle $\times$ saddle) for the third (orange) and fifth (blue) families while the remaining ones are of type B_2 (saddle $\times$ centre).

The periodic orbits associated to the other three equilibria can be obtained using the symmetries III_N, IV_N and V_N. They are shown in Fig. 4.17. We stress that there is no qualitative difference with other values of β from the one shown here.

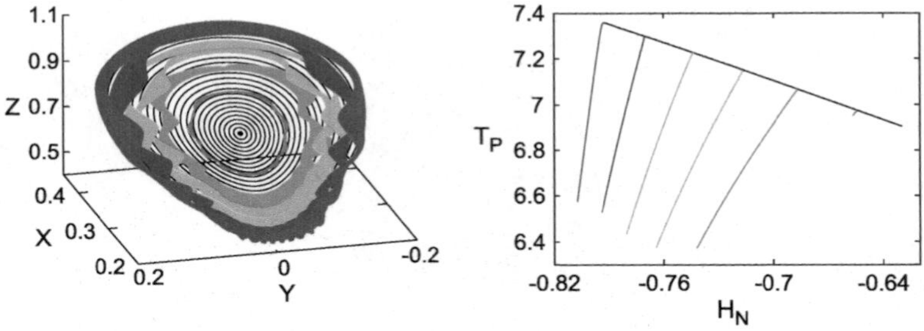

Fig. 4.16 The six bifurcated families of periodic orbits born from saddle-node bifurcations, together with the family from which they emanate (black). The periodic orbits (left) and the corresponding characteristic curves (right) are shown in the same colors

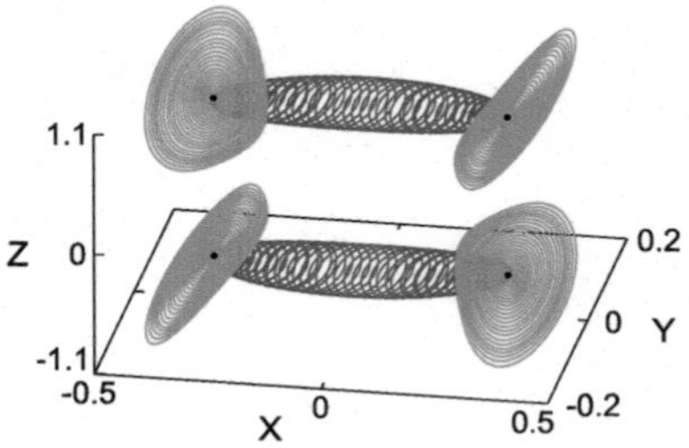

Fig. 4.17 For $\beta = 2$, the six families of periodic orbits emerging from the four equilibria of type 2_N. The orbits have been represented using the ones computed for $X > 0$, $Z > 0$ and the III_N, IV_N and V_N symmetries

4.3.4.3 Periodic Orbits Around the 3_N Equilibria

There are only two equilibrium points of type 3_N, and their coordinates are $(\pm(1/3)^{1/3}, 0, 0)$. For any value of β, the Jacobian at these points has four purely imaginary eigenvalues, $\pm\lambda_1 i$ and $\pm\lambda_2 i$, purely imaginary and two real eigenvalues, $\pm\lambda_3$. As a consequence, there are two families of periodic orbits in the centre manifold associated to each point. Taking $\beta = 2$ for illustration purposes (again the qualitative behaviour of the periodic orbits is independent of β), we have $\lambda_1 = 7.6460$, $\lambda_2 = 3.1623$ and $\lambda_3 = 0.6796$, so the periods of the two families tend to $2\pi/\lambda_1 = 0.821761092$ and $2\pi/\lambda_2 = 1.986903617$, respectively. Some orbits of both families, together with the characteristic curves, are shown in Fig. 4.18.

The orbits of the family associated with $\lambda_1 = 7.6460$ are planar. Close to the equilibrium point, the orbits are unstable. At $H_n = 3.827218$ (the first vertical dashed line) a saddle-node bifurcation is detected and the family transits from type B_2 (unstable) to B_3 (stable). As it is shown in Fig. 4.18, close to its end the orbits become infinitesimal oscillations around the origin, where the family ends.

The orbits of the family associated to $\lambda_2 = 3.1623$ are 3-dimensional and reach a planar orbit ($Z = 0$) with $H_n = 3.868084$, travelled four times, which is marked with an A in the left plots of the two bottom lines of Fig. 4.18. The orbits of this piece of the family go from type B_2 (unstable) to type B_3 (stable) at $H_n = 3.569571$, where there is a saddle-node bifurcation. Very close to the end of the family ($H_n = 3.868084$), the type changes from B_3 to B_4, and then back to B_3 through two Krein bifurcations.

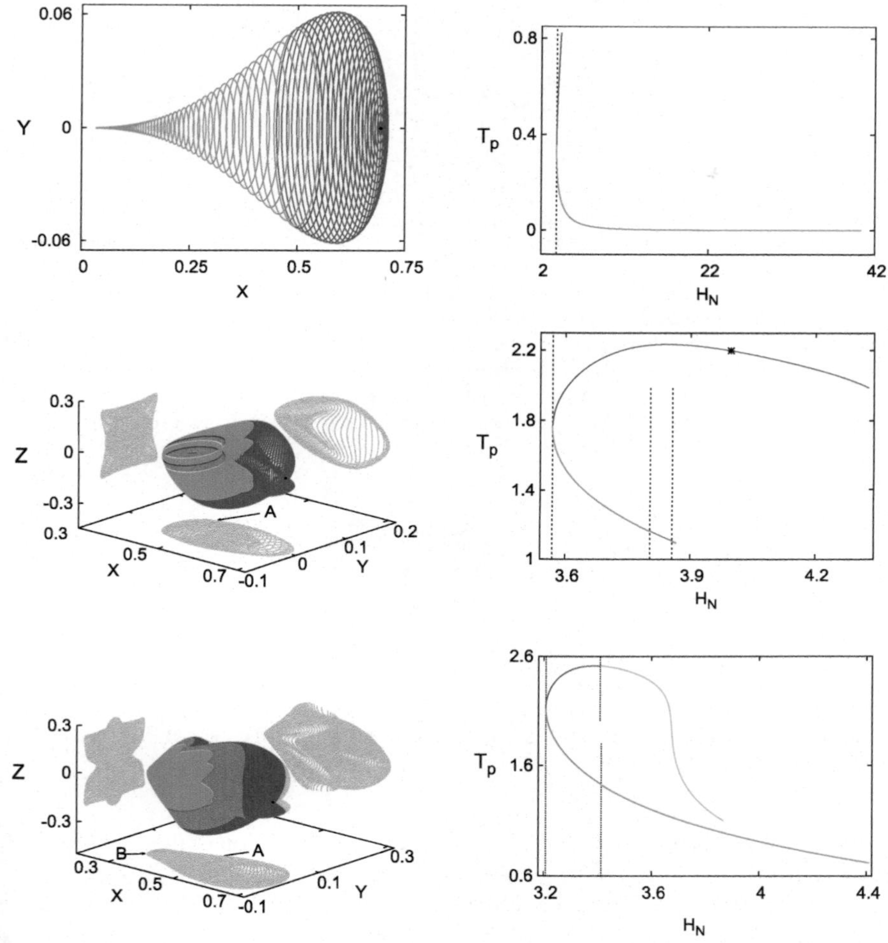

Fig. 4.18 The two families of periodic orbits emanating from the 3_N equilibrium point with $X > 0$. The top row shows some orbits of the family associated with the eigenvalue λ_1 together with the characteristic curve. All orbits of this family are planar; those of type B_3 are in green (stable orbits), and those of type B_2 in purple (unstable orbits). The second and third lines show the results for the family associated to the eigenvalue λ_2. The evolution of the family up to the planar orbit with $Z = 0$ (label A) is displayed in the middle line, the orbits in green are the stable ones and the colors indicate the type evolution of the orbits, that follows the sequence B_2 (purple) $\rightarrow$ B_3 (green) $\rightarrow$ B_4 (red) $\rightarrow$ B_3 (green). The bottom line shows the orbits obtained from the orbit with label A up to the orbit with label B with the same color criteria as above. In these last orbits the type transitions follow the sequence: B_2 (orange) $\rightarrow$ B_1 (blue) $\rightarrow$ B_2 (magenta) $\rightarrow$ B_3 (green)

Some orbits of this family, with a different number of loops after each complete revolution, are shown in Fig. 4.19.

Starting at the planar orbit with $H_n = 3.868084$, we have proceed with the continuation procedure and found that at A there is a turning point along the family,

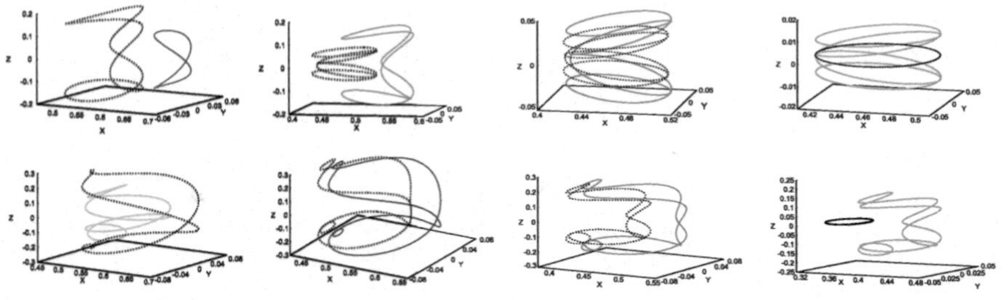

Fig. 4.19 Evolution of orbits of the family associated with λ_2 around the 3_N equilibrium point with $X > 0$. The first and second rows are in correspondence to the middle and bottom rows in Fig. 4.18 respectively. Each subplot displays the orbits of one type up to the next bifurcation (or termination in the last one). The representation of the orbits uses the same color criteria as in Fig. 4.18 and the black ones correspond to bifurcations (dashed line) or termination (solid line). The loops within one revolution go from two to four in the first row and from four to six in the second one

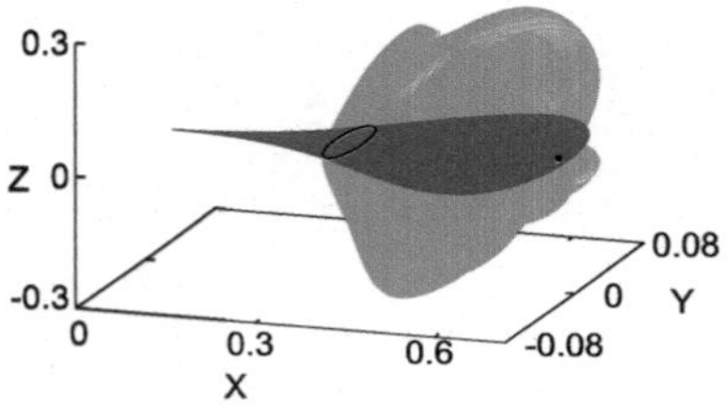

Fig. 4.20 3D representation of the two families of periodic orbits emanating from the equilibrium point 3_N ($X > 0$). The two orbits corresponding to the ones with labels A and B in the bottom line of Fig. 4.18 are displayed in black

where the type changes from B_3 to B_2. Then the type evolution of the orbits follows the sequence: $B_2 \rightarrow B_1 \rightarrow B_2 \rightarrow B_3$, where only the final type B_3 (green) is stable; the termination occurs at the orbit indicated with the label B ($H_n = 4.410427$), where the family meets an orbit of the planar family associated to λ_1 travelled six times. Along this branch three saddle-node bifurcations are detected when the three transitions happen. They are indicated with vertical dashed lines in the characteristic curve. The orbits of this branch and its characteristic curve are shown in the bottom line of Fig. 4.18. In the energy level marked with black dots inside the characteristic curves shown in Fig. 4.18 there are periodic orbits with centre part being used for the computation of 2D tori in the following section.

The orbits of the second family (associated with λ_2) have complicated shapes moreover seven bifurcations have been detected along this family before its end. Figure 4.19 shows the evolution of the orbits of each topology type up to a bifurcation. A joint representation of the two families of periodic orbits is given in Fig. 4.20 that clearly shows the two intersections corresponding to the two orbits labelled as A and B in Fig. 4.18.

We have also computed a new family born from a saddle-node bifurcation along the branch in the middle line of Fig. 4.18. The orbits of this family and the associated

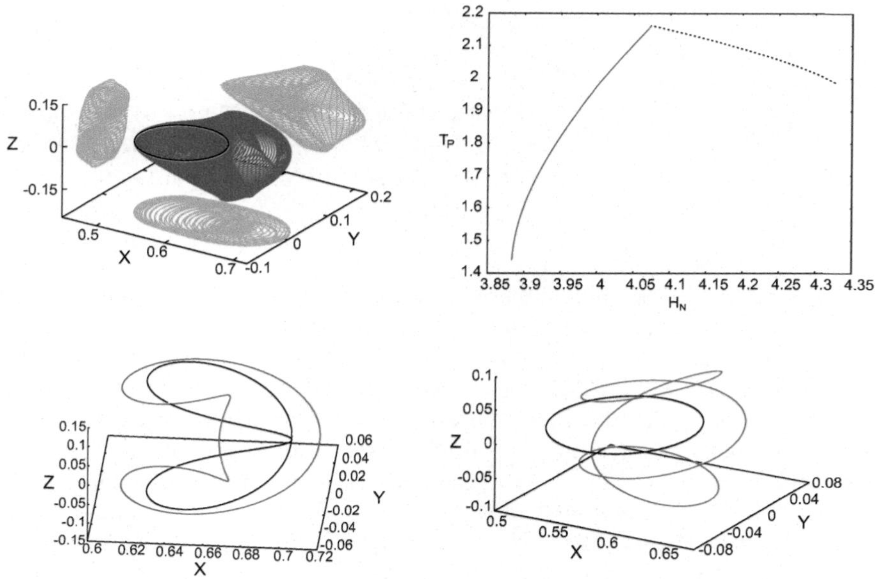

Fig. 4.21 Bifurcated family born at the first saddle-node bifurcation of the branch in the middle row of Fig. 4.18. The first row shows the orbits (left) and the associated characteristic curve (right). The dashed black line in the right hand side plot corresponds to the original branch before the first saddle-node bifurcation. The second row displays the orbit evolution; the orbits in black are the starting one (left) and and the final one (right)

characteristic curve are shown in Fig. 4.21. They are unstable and of type saddle $\times$ centre (B_2). The family ends at an orbit associated with the planar family with $H_n = 3.883665$ travelled three times.

4.3.4.4 Quasi-periodic Orbits Around Periodic Orbits

In this section we show some of the results obtained in the computation of families of quasi-periodic orbits ($2D$ invariant tori) using the procedure explained in Sect. 4.3.2.

We have computed three families of $2D$ tori: one family is associated with periodic orbits around the equilibrium 1_N, and the other two are associated with periodic orbits around the equilibrium 3_N. All three families are iso-energetic, so they are computed by keeping the energy H fixed and varying the rotation number and the return time (ρ, T) during the continuation. Analogously to what has been done with periodic orbits, the curve (ρ, T) can be seen as their characteristic curve.

The (ρ, T, H) values of the three periodic orbits where we start the computations are listed in Table 4.5 as well as their initial conditions. These three periodic orbits are the ones displayed with black dots in their associated characteristic curves in Figs. 4.12 and 4.18.

Table 4.5 Values of (ρ, T, H) and initial conditions of the periodic orbits around which we have computed the isoenergetic families of 2D tori

Label	ρ	T_p	H_N	χ_0 $(X_0, Y_0, Z_0, X_0', Y_0', Z_0')$
1_{N1}	1.454749	2.196629	−1.798693	(0.0, 0.932165, 0.701220, 0.460454, 0.0, 0.0)
3_{N1}	0.745569	2.197733	3.996198	(0.699132, 0.0, 0.0, 0.0, −0.158072, 0.553048)
3_{N2}	0.790994	0.903410	4.008587	(0.338497, 0.0, 0.059964, −0.059654, 1.358685, 0.350506)

The results corresponding to the isoenergetic family of tori around the periodic orbit with label 1_{N1} ($H_N = -1.798693$) are displayed in Fig. 4.22. The invariant curves φ associated to the tori (top left plot of the figure) start with a value of the rotation number $\rho = 1.454749$, and are the small circles in the middle of the set of invariant tori. As the family evolves the size of the torus becomes larger and the shape of the corresponding invariant curve turns from simple circles to more complicated and bended curves. The family has been computed up to values of ρ close to the resonance $\rho = 2\pi\frac{15}{64}$, indicated as dashed line in the top right hand side plot of Fig. 4.22 (characteristic curve). The same figure also shows a 3D representation of some of the computed tori together with the invariant curve used for its computation.

Figure 4.23 shows the results obtained for the isoenergetic family of tori around the periodic orbit with label 3_{N1} ($H_N = 3.996198$). This family has been computed up to the resonance with $\rho = \frac{2\pi}{29}$. As it is clearly seen at the invariant curves, the shape becomes more complex as the tori increase in size and, as a consequence, the number of frequencies N_f required for their accurate computation varies along the family. To show this fact, we have plotted the invariant curves in different colors according to the values of N_f. The continuation procedure terminates when we approach the 1/29 resonance, which corresponds to the maximum allowed value of $N_f = 128$.

Figure 4.24 shows the results obtained for the isoenergetic family of tori around the periodic orbit with label 3_{N3} ($H_N = 4.008587$). Recall that this orbit belongs to the same family as the periodic orbit 3_{N1}. This family has been computed up to the resonance corresponding to $\rho = \frac{2\pi}{27}$, when the number of frequencies required for the computation of the invariant curve φ is greater that the maximum allowed value $N_f = 128$.

4.3.5 The Radial Case

In the radial case, and for all values of β, we have three kinds of equilibria: 1_R, 2_R and 3_R. The two 1_R equilibria are of type centre × saddle × saddle, which means that there is one one-parameter family of periodic orbits around each of them. The second kind of equilibria is of type centre × centre × saddle, that means we have two families of periodic orbits associated to all the four 2_R points. The type of the third kind of equilibria depends on the value of β, which can be either centre × centre × centre or centre × saddle × saddle.

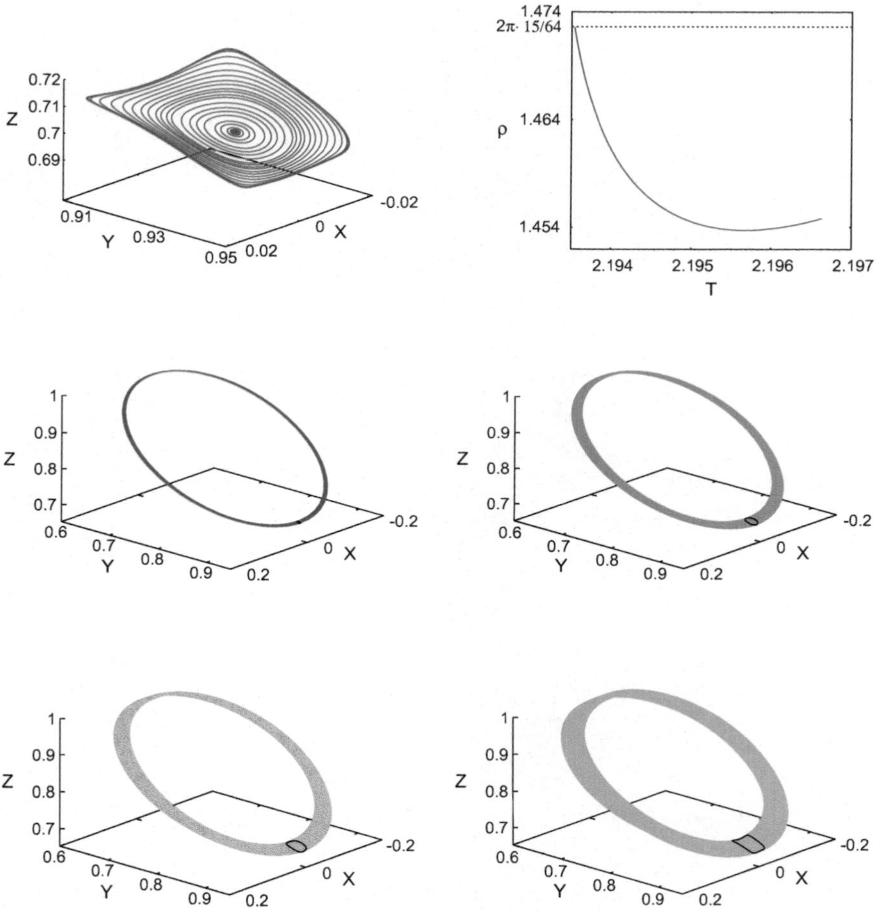

Fig. 4.22 For $H_N = -1.798693$, invariant curves φ associated to the isoenergetic family of 2D tori around the periodic orbit with label 1_{N1} (top left plot). The top right plot shows the rotation number ρ in front of return time T of the isoenergetic family of tori. The family has been computed up to values of ρ close to the resonance $15/64$. The two bottom rows show some of the tori of the family. In all the plots we have included the invariant curves φ used for their computation (in black)

For the computation of the families of periodic orbits we only consider some values of β. In all cases the values are chosen such that the dimension of the associated centre manifold is maximal. We note that once the stability of the equilibrium point has been fixed, the qualitative behaviour of the periodic orbits emanating from the point is independent of β, although the size of the orbits changes with the value of this parameter.

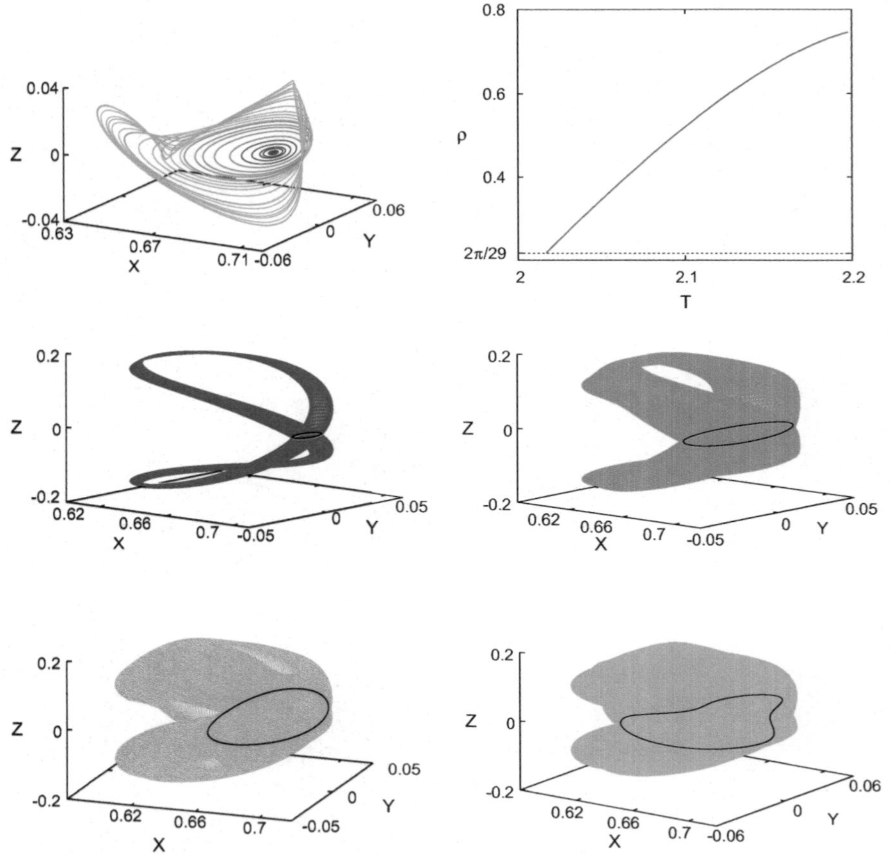

Fig. 4.23 For $H_N = 3.996198$, invariant curves φ associated to the isoenergetic family of 2D tori around the periodic orbit with label 3_{N1} (top left plot). The different colors of the curves correspond to the number of frequencies required for their computation (≤ 16 magenta, ≤ 32 cyan, ≤ 64 orange and ≤ 128 blue).The top right plot shows the rotation number ρ in front of return time T of the isoenergetic family of tori. The family has been computed up to values of ρ close to the resonance $1/29$. The two bottom rows show some of the tori of the family. In all plots we have included the invariant curves φ used for their computation (in black)

4.3.5.1 Periodic Orbits Around the 1_R Equilibria

There are two equilibria of type 1_R located at $(0,\ 0,\ \pm 1)$. We will focus on the one with $Z = 1$. The family of periodic orbits obtained with $\beta = 2$ is displayed in Fig. 4.25. We stress that the qualitative behaviour of the periodic orbits is independent of β. The family associated to $(0,\ 0,\ -1)$ can be obtained by applying the III_R symmetry w.r.t. the origin.

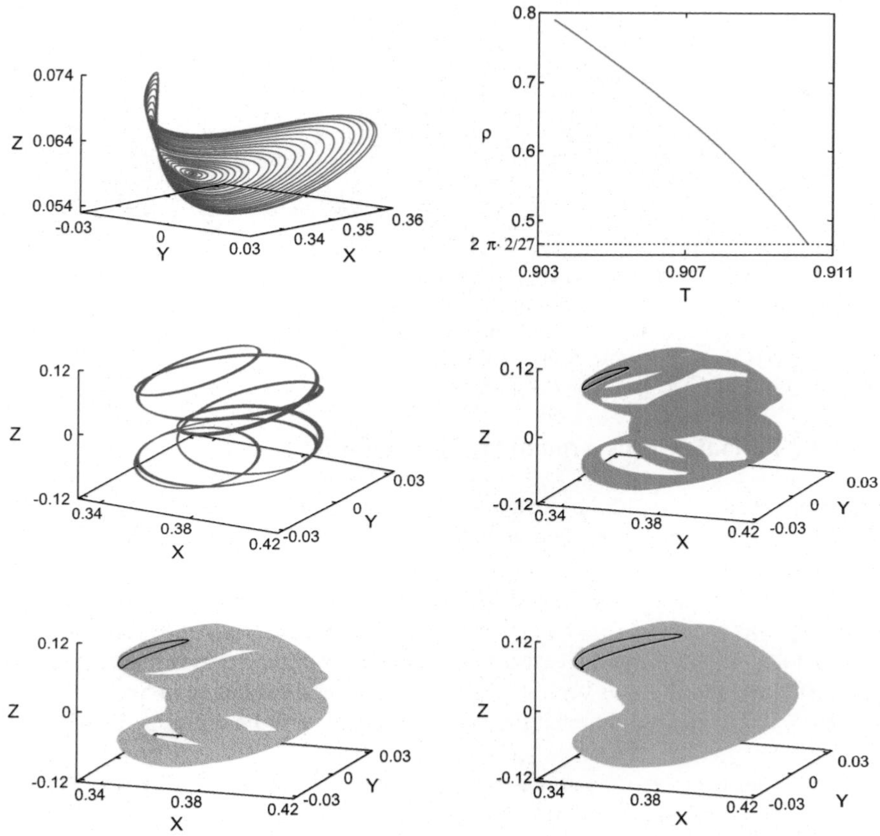

Fig. 4.24 For $H_N = 4.008587$, invariant curves associated to the isoenergetic family of 2D tori around the periodic orbit with label 3_{N2} (top left plot). The top right plot shows the rotation number ρ in front of return time T of the isoenergetic family of tori. The family has been computed up to values of ρ close to the resonance $2/27$. The two bottom rows show some of the tori of the family. In all plots we have included the invariant curves φ used for their computation (in black)

The orbits in this family are all unstable and the type of the multipliers goes from B_1 to B_4, with the transition at $H_R = -3.159252$. No bifurcation is detected at this transition since the stability does not change. Analogous to the 1_N case with $\beta = -2$ (see Fig. 4.13), the orbits grow along the Z direction as the energy decreases and the period tends to an upper limit value close to π.

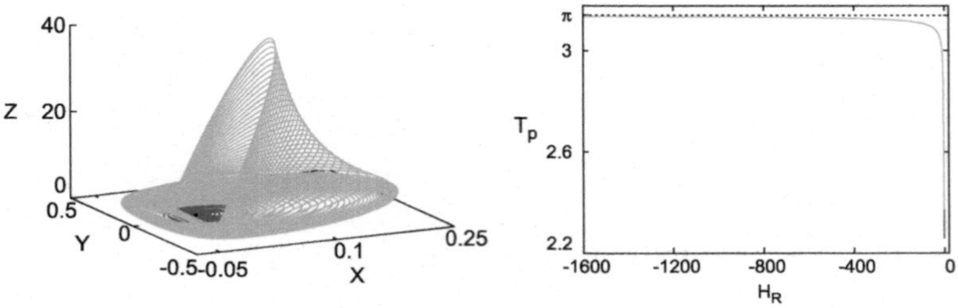

Fig. 4.25 For $\beta = 2$, family of periodic orbits emanating from the equilibrium $(0, 0, 1)$ of type 1_R $(Z = 1)$. The left plot shows the orbits of the family and the right one the associated characteristic curve. The colors in both figures indicate different configurations of the multipliers, along which the family goes from B_1 (purple) to B_4 (cyan)

4.3.5.2 Periodic Orbits Around the 2_R Equilibria

There are four 2_R equilibria located at $(\pm((2/(9\sqrt{3}))^{1/3}, \pm\sqrt{2}(2/(9\sqrt{3}))^{1/3}, 0)$. We will mainly focus on the one with $X > 0$ and $Y > 0$. For $\beta = 2$, the associated eigenvalues are $\lambda_1 = \pm 4.3528i$, $\lambda_2 = \pm 1.2784i$, and $\lambda_3 = \pm 0.7624$, so there is one family of periodic orbits associated with λ_1 and another one with λ_2. Since the equilibrium points are in the $X - Y$ plane, we cannot determine symmetric periodic orbits (recall Table 4.1 for the radial case). The computation of the non-symmetric periodic orbits of these families has been done using the method explained in Sect. 4.3.1 taking $Z_0 = 0$. The results obtained for both families of periodic orbits are displayed in Fig. 4.26.

Analogous to some previous cases, such as the 1_N case with $\beta > 0$ or the 2_N case, the family associated to λ_1 terminates at the equilibrium point symmetric w.r.t. the Y axis. In fact, the family emanating from $((2/(9\sqrt{3}))^{1/3}, +\sqrt{2}(2/(9\sqrt{3}))^{1/3}, 0)$ and the one $((2/(9\sqrt{3}))^{1/3}, -\sqrt{2}(2/(9\sqrt{3}))^{1/3}, 0)$ are the same one, due to the fact that the orbits around one point are symmetric with the ones around the other point w.r.t the Y axis (II_R symmetry). As a consequence, the characteristic curve is, in fact, the one displayed in the top row of Fig. 4.26 travelled twice. We note that the most left point of the curve corresponds to the orbit connecting the two families, which is a degenerated centre. It is the black curve on the left hand side figure that has been indicated with the vertical line in the characteristic curve.

All the orbits of this family are unstable, and for the branch starting at $H_R = 2.289428$ and ending at $H_R = 1.271177$ (the connecting orbit), the transition history of the configuration type follows: $B_2 \to B_1 \to B_2$, with two periodic-doubling bifurcations at the two transitions. For the symmetric branch, the transitions and bifurcations follow a reverse pattern.

The orbits of the family associated with λ_2, together with the associated characteristic curve, are displayed in the bottom row of Fig. 4.26. The orbits of this family are unstable and the transitions follow the sequence: $B_2 \to B_1 \to B_2 \to B_1 \to B_4 \to B_1$. Two period-doubling bifurcations occur at the first and third transitions, and a

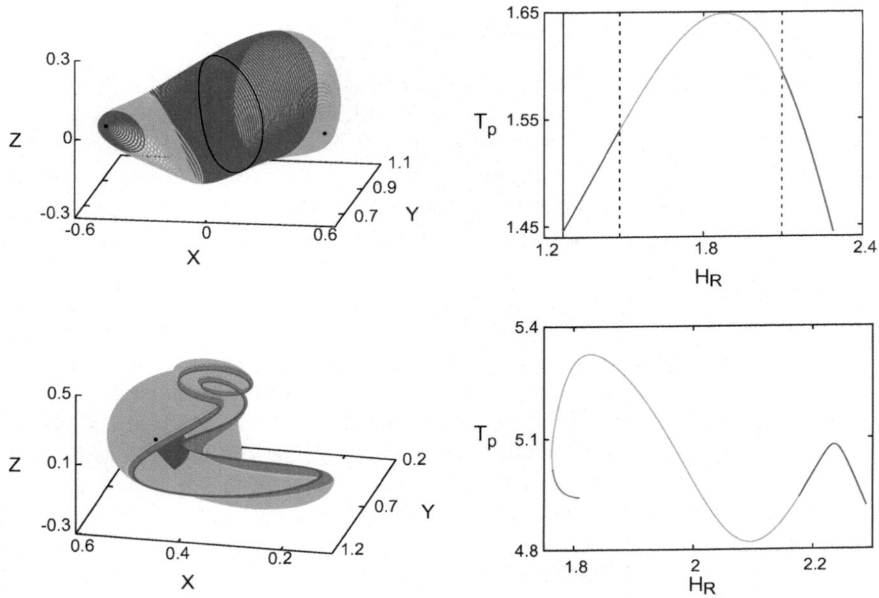

Fig. 4.26 The two families of periodic orbits emanating from the 2_R equilibria with $Y > 0$. The first row shows the results for the family associated to λ_1. All orbits are unstable and the transition of their type follows: B_2 (purple) $\rightarrow$ B_1 (cyan) $\rightarrow$ B_2 (purple). Two bifurcations occur at the transition points, and indicated as vertical lines. The black line on the left, which corresponds to the black orbit displayed in left frame, indicates the place where the two branches emanating from the two 2_R equilibria meet and a saddle-node bifurcation occurs. The bottom row corresponds to the family associated to λ_2. All orbits are unstable and their types go through B_2 (purple) $\rightarrow$ B_1 (cyan) $\rightarrow$ B_2 (orange) $\rightarrow$ B_1(cyan) $\rightarrow$ B_4 (magenta) $\rightarrow$ B_1 (blue)

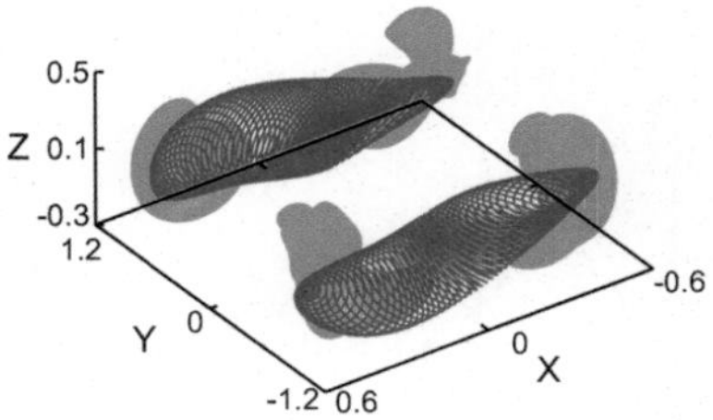

Fig. 4.27 All the families of periodic orbits emanating from 2_R equilibria when $\beta = 2$. The displayed orbits have been obtained by applying I_R, 2_R and 3_R symmetries from the ones computed for $X > 0$ and $Y > 0$

saddle-node bifurcation occurs at the second one. No bifurcation happens at the last two transitions since there are no changes in stability of the orbits.

The periodic orbits around the remaining equilibria can be obtained applying the symmetries I_R, II_R or III_R. The resulting families are displayed in Fig. 4.27, together with the two connections detected between each two equilibria, which are symmetric w.r.t. the Y axis.

4.3.5.3 Periodic Orbits Around the 3_R Equilibria

There are four equilibrium points with label 3_R, located at $(\pm(1/(4\sqrt{2}))^{1/3}$, 0, $\pm(1/(4\sqrt{2}))^{1/6})$. We recall that for the two equilibria with $Z = X$ and $\beta \in (-\infty, -0.9516) \cap (3.4525, \infty)$, and also for the equilibria with $Z = -X$ and $\beta \in (-\infty, -3.4525) \cap (0.9516, \infty)$, the maximum dimension of the associated centre manifold is six. In both cases we have three families of periodic orbits around each equilibrium point. For the computation of the periodic orbits we have used the type I_R symmetry w.r.t. the $X - Z$ plane.

For the numerical results that follow, we have used $\beta = 6$ and considered only the two equilibrium points with $X = Z > 0$. The associated eigenvalues for this value of β are $\lambda_1 = \pm 29.6282i$, $\lambda_2 = \pm 1.4666i$ and $\lambda_3 = \pm 0.1381i$. Figure 4.28 shows the results obtained for the three families of periodic orbits. The first family goes through the type evolution $B_3 \to B_2 \to B_3$ with two saddle-node bifurcations and terminates in infinitesimal oscillations around the origin. The second one is stable (B_3) and terminates at a degenerated centre, while the third family goes from B_3 to B_2, after a saddle-node bifurcation at $H_R = 1.917373$, and terminates at a degenerate saddle.

A connection between the family associated with λ_1 and its image w.r.t. the origin (Type III_R symmetry) has been detected. The connecting family is shown in Fig. 4.29.

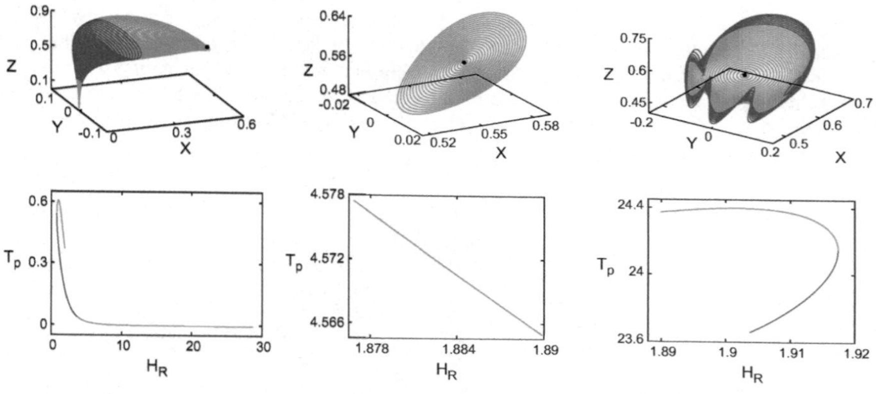

Fig. 4.28 The three families of periodic orbits emanating from the 3_R equilibria, with $X = Z > 0$, associated with λ_1 (left), λ_2 (middle) and λ_3 (right). The different colors, of both the orbits and the characteristic curves, indicate different configurations of the multipliers associated with the periodic orbits (the stable ones in green and the unstable ones in other colors). In the family associated with λ_1 the orbits are stable (B_3) before a saddle-node bifurcation occurs at $H_R = 0.755442$ and turn into unstable (B_2). At $H_R = 3.710629$ a second saddle-node bifurcation occurs and the family remains stable (B_3) until its termination. The orbits of the family associated with λ_2 are stable (B_3) and terminate at a degenerate centre at $H_R = 1.876817$. The orbits of the family associated with λ_3 go from B_3 (stable) to B_2 (unstable). The transition happens at a saddle-node bifurcation at $H_R = 1.917373$ and the family terminates at a degenerate saddle

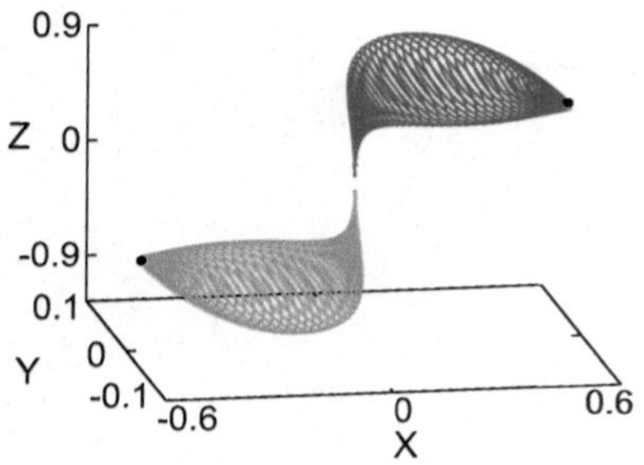

Fig. 4.29 Connection between the family associated with λ_1 and its image w.r.t. the origin (Type III_R symmetry)

The two equilibrium points with $X = -Z$, for $\beta \in (-\infty, 13.4525) \cup (0.9516, +\infty)$ have also a 6D centre manifold associated. Fig. 4.30 shows the three families of periodic orbits obtained in this case when $\beta = 6$. The first family (left) goes from B_3 (stable) to B_4 (unstable) type after a Krein bifurcation at $H_R = -2.046653$. Analogous to what happens to the family associated to the 1_R equilibrium point (Fig. 4.25), the size of the orbits becomes rather large as the energy decreases, while the period tends to an upper limit close to π. The orbits of the second family (middle) are stable and of type B_3. This family terminates at a degenerated centre when $H_R = 1.822400$. The orbits of the third family (right) are stable and terminates at a degenerated centre at $H_R = 2.063967$. In the second and the third families there is a monotonic relation between the period and the energy, as is clearly shown in their associated characteristic curves.

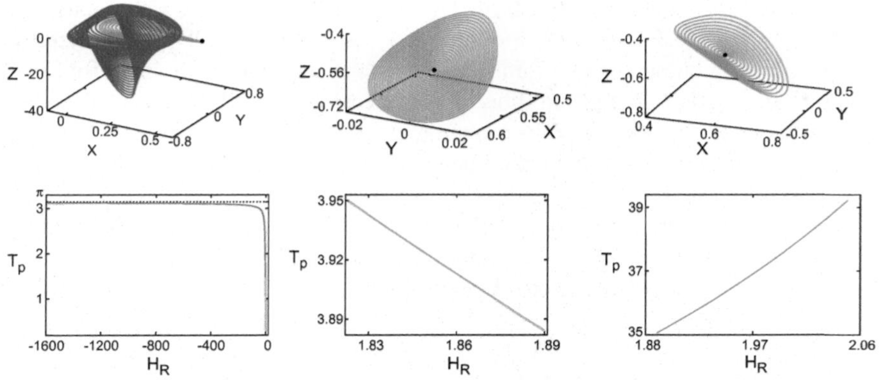

Fig. 4.30 The three families emanating from equilibria 3_R with $Z = -X$, $X > 0$ and $\beta = 6$. The family associated with λ_1 (left) is stable (B_3) up to $H_R = -2.046653$ where a Krein bifurcation occurs. After this, the orbits become complex unstable (B_4). The orbits of the other two families are stable (B_3). The second family terminates at a degenerate centre at $H_R = 1.822400$, while the third family ends at a degenerated saddle at $H_R = 2.063967$

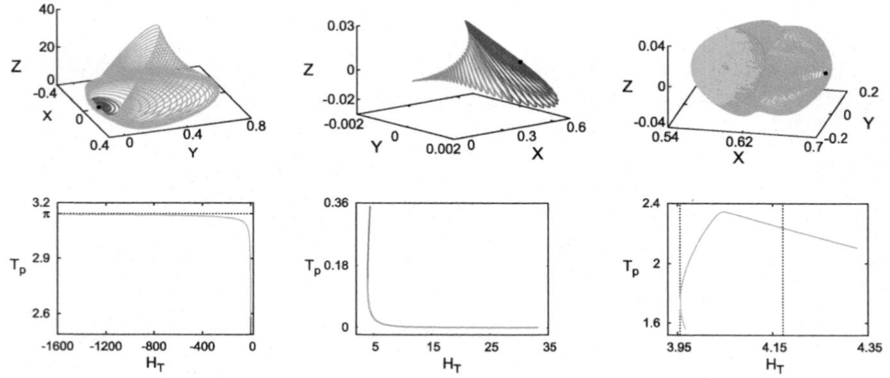

Fig. 4.31 The three families of periodic orbits in the tangential case. The orbits of the family emanating from (0, 0, +1) (left) are unstable and transit from B_1 (purple) to B_4 (cyan). The two families associated with $((1/3)^{1/3}, 0, 0)$ are shown in the remaining columns. The orbits of one of the families transit from B_2 (purple) to B_3 (green) type, after a saddle-node bifurcation at $H_T = 3.914646$. The orbits of the other family transit from B_2 (cyan)to B_1 (orange) when a saddle-node bifurcation occurs at $H_T = 3.955519$. This family ends at a degenerated saddle at $H_T = 3.965645$. At the saddle-node bifurcation indicated with the right-hand side vertical line, another family is born (see Fig. 4.33)

4.3.6 The Tangential Case

We recall that in the tangential case there are two kinds of equilibrium points: two 1_T equilibria located at $(0, 0, \pm 1)$, and two 2_T equilibria with coordinates $(\pm(1/3)^{1/3}, 0, 0)$. For any value of β, the 1_T points have a 2D centre manifold which embeds a one-parameter family of periodic orbits. The centre manifold associated to the second kind of equilibria has dimension four which means that there are two families of periodic orbits around them. As in the preceding cases, we will focus on the equilibrium point in the first quadrant. The results for $\beta = 6$, are shown in Fig. 4.31.

The orbits associated with $(0, 0, +1)$ are all unstable and transit from B_4 (complex saddle) to B_1 (real saddle) without any bifurcation in between. The continuation procedure is finished when the orbits become very large ($|Z| > 40$). Analogously to

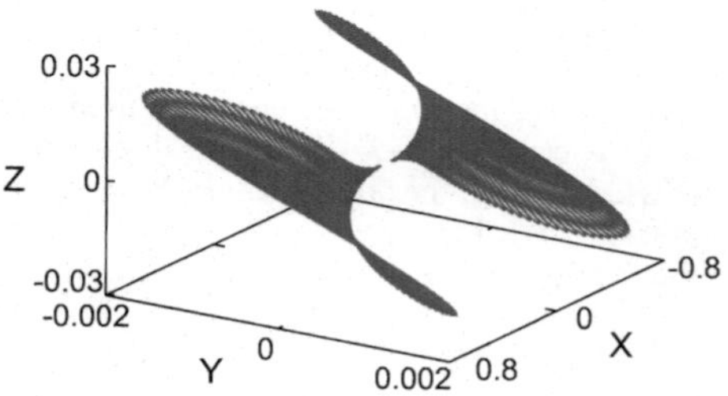

Fig. 4.32 Connection between the families associated to $((1/3)^{1/3}, 0, 0)$ and $(-(1/3)^{1/3}, 0, 0)$

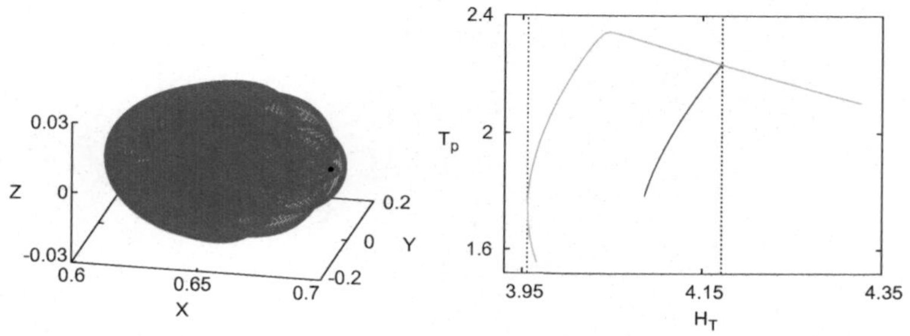

Fig. 4.33 Bifurcated family emanating from the saddle-node bifurcation appearing in the second family associated to $((1/3)^{1/3},\ 0,\ 0)$

the families shown in Figs. 4.13 and 4.30 (top row), an upper limit close to π for the period exists, even the size of the orbit grows rather large.

The two families associated with $((1/3)^{1/3},\ 0,\ 0)$ are also shown in Fig. 4.31. The orbits of the first family associated to this point are of type B_2 (unstable) and transit to B_3 (stable) after a saddle-node bifurcation at $H_T = 3.914646$. The last orbits of the family are infinitesimal oscillations around the origin, where the family ends. A connection at the origin is detected between the families associated to $((1/3)^{1/3},\ 0,\ 0)$ and $(-(1/3)^{1/3},\ 0,\ 0)$, which are symmetric w.r.t. the origin (see Fig. 4.32).

The orbits of the second family associated to $((1/3)^{1/3},\ 0,\ 0)$ are unstable and transit from B_2 to B_1 type at $H_T = 4.326748$. The family ends as a degenerated saddle at $H_T = 3.965645$, which corresponds to the final point of the characteristic curve. Two saddle-node bifurcations have been detected, the first one at $H_T = 4.170688$, without any change of stability, and the second at $H_T = 3.955519$. We have computed the bifurcated family born at the second saddle-node bifurcation (see Fig. 4.33).

4.4 Formation Flying Configuration Design

4.4.1 Formation Flying Configuration Using Equilibrium Points

As stated in Sect. 4.2.4, for each of the three cases considered, there are several kinds of equilibrium points, which are stationary points relative to the leader in the LHLV coordinate system, and provide straightforward locations for formation flying configuration. Several configurations are shown in Fig.4.34, in each configuration one kind of equilibria of the same energy level are used and marked in the same color.

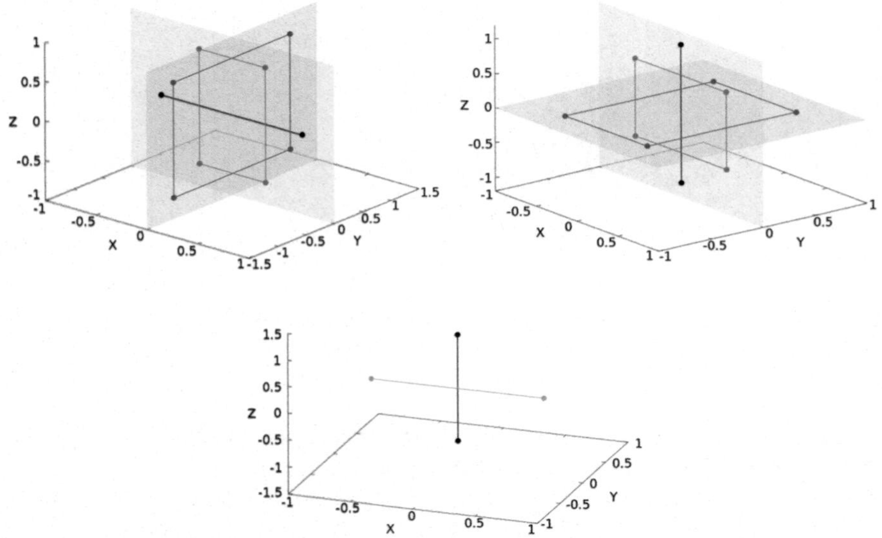

Fig. 4.34 Several formation flying configurations using the equilibria as the nominal location, in normal case (**top left**), radial case (**top right**) and tangential case (**bottom**). In each configuration, the satellites (round points) are deployed at one kind of equilibria, which are indicated in different colors and by segments connecting the adjacent two satellites

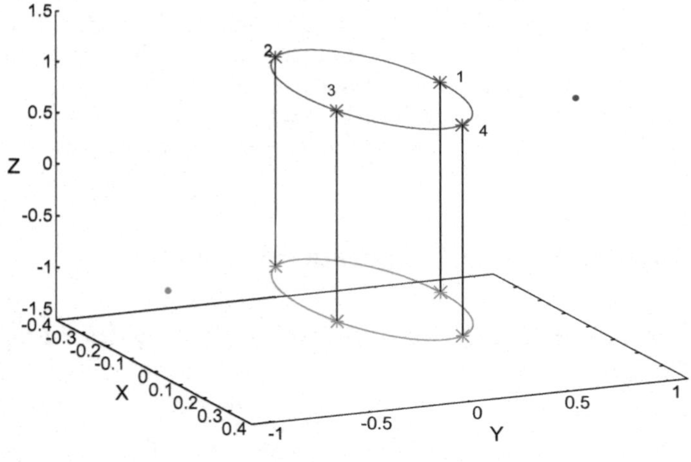

Fig. 4.35 The two-satellite formation flying configuration using the middle green orbits in Fig. 4.14. We display the configuration segment that connects the two satellites (stars) at epoches $T/4 * (i - 1)$, $(i = 1, 2, 3, 4)$ with T as the period. The two round points are the two equilibria 1_N symmetric w.r.t. the origin

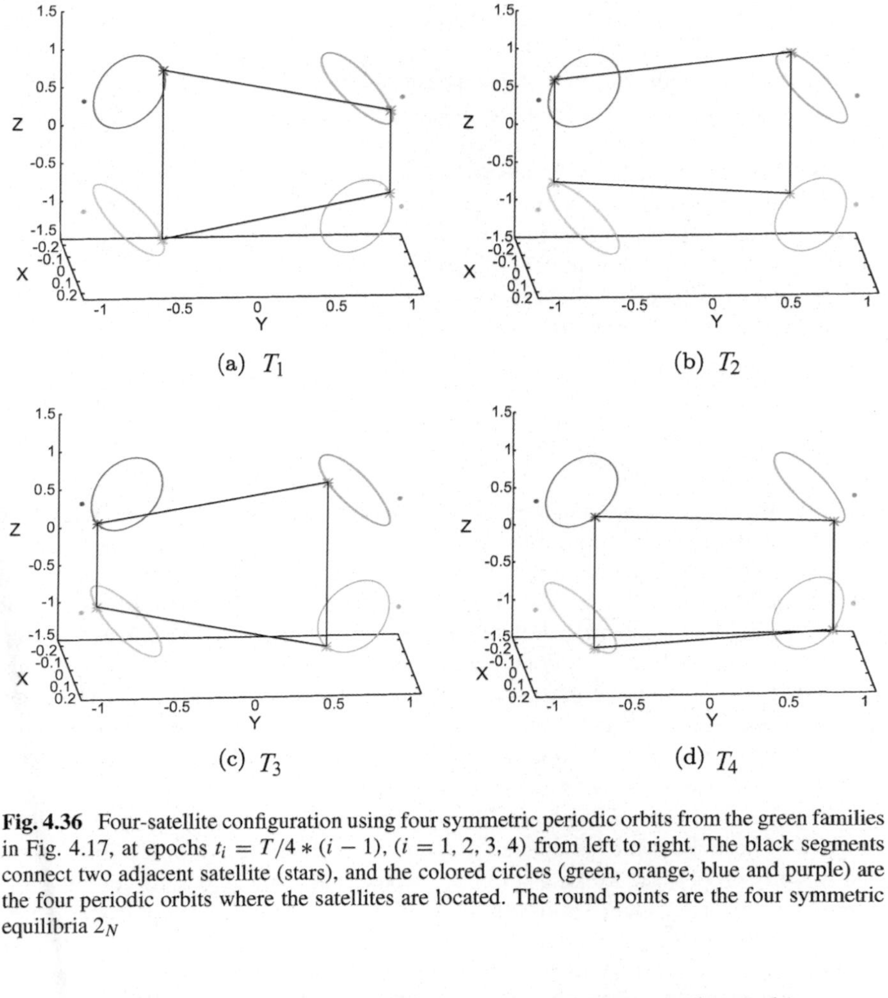

Fig. 4.36 Four-satellite configuration using four symmetric periodic orbits from the green families in Fig. 4.17, at epochs $t_i = T/4 * (i-1)$, $(i = 1, 2, 3, 4)$ from left to right. The black segments connect two adjacent satellite (stars), and the colored circles (green, orange, blue and purple) are the four periodic orbits where the satellites are located. The round points are the four symmetric equilibria 2_N

4.4.2 *Formation Flying Configuration Using Periodic Orbits*

The other option for formation configuration is to use the symmetric periodic orbits as the nominal orbits, which have the same period and stability behaviour.

Using the two green periodic orbits in Fig.4.14 that are "almost" parallel to $X - Y$ plane, we design a two-satellite configuration, with one satellite placed on one orbit. For missions that require the distance between two satellites be constant, these two periodic orbits are perfect nominal orbits. We display the configuration at four epochs during one period, see Fig. 4.35, the sense of satellites' motion is counter-clockwise.

Considering the second family of periodic orbits around the four equilibrium points with label 2_N(see green ones in Fig. 4.17), a four-satellite configuration is designed. Note that these orbits are obtained using the symmetries of type III_N, IV_N

and V_N, which have the same time direction, and move in counter-clockwise sense (Fig. 4.36).

Let us recall that, the real unit of distance is determined by the value of the charge-to-mass ratio $\frac{q}{m}$, β, and the field moment B_0, while the unit of time is independent on these parameters. The magnetic strength B_0, as well as β, is considered fixed once the HTSC is designed, but the value of $\frac{q}{m}$ remains adjustable. As a consequence, a configuration with more satellites can be designed using any of the configuration above, but with different charge on extra satellites. For instance, a 12-satellite (of the same mass) configuration using equilibria 1_N, with the charge of each four satellites as q_1, q_2 and q_3, respectively, where we have $q_1 = 2q_2 = 4q_3$; the final configuration will be three rectangles of ratio $1 : 2^{\frac{1}{3}} : 3^{\frac{1}{3}}$.

4.5 Conclusions

In this chapter, we have studied the relative dynamics of a charged spacecraft moving around a leader spacecraft provided with a rotating magnetic dipole. The leader is assumed to move following circular orbits around a central body (for instance, the Earth). Three basic cases for the orientation of the dipole have been considered: normal, radial and tangential. For the equations associated to the three resulting dynamical systems, we have done an exhaustive analysis of their equilibria, and their stability behaviour with respect to the parameter β (quotient between the mean motion of the leader and the angular velocity of the dipole). We have also explored the admissible regions of motion in the phase space, by analysing the topology of ZVS when the energy goes through the critical values at the equilibria.

The phase space around each equilibria has been carefully explored by computing families of periodic orbits associated to them, including bifurcations, terminations, and stability properties. All kinds of bifurcations: saddle-node, period-doubling and Krein, have been detected. Several families of bifurcated families born from the first kind of bifurcation have been computed. Furthermore, a parameterisation method, based on stroboscopic map method, has been developed and applied to compute 2D invariant tori around elliptic periodic orbits. The relation between the Poincaré map method and the stroboscopic map method has been established in terms of the return time. Several families of 2D tori have been computed, using the linear flow around the elliptic periodic orbits as initial seed.

To demonstrate the potential application of the model, several formation flying configurations have been designed using the equilibrium points, and periodic orbits as the nominal orbits.

The models considered in the chapter have great potential use for future space mission applications. The overall work provides an exhaustive catalog of periodic and quasi-periodic orbits for selecting candidates as nominal trajectories for formation flying, long time hovering missions, etc.

References

1. E. Anderson, Z. Bai et al., *LAPACK Users' Guide*, 3rd edn., Society for Industrial and Applied Mathematics (1999)
2. J. Atchison, M. Peck, Lorentz-augmented Jovian orbit insertion. J. Guid. Control Dyn. **32**(2), 418–425 (2009)
3. N. Baresi, Z.P. Olikara, D.J. Scheeres, Survey of numerical methods for computing quasi-periodic tori in Astrodynamics, in *26th AAS/AIAA Space Flight Mechanics Meeting, Napa, CA* (2016)
4. B. Baudouy, Heat transfer and cooling techniques at low temperature (2015), arXiv:1501.07153
5. L.S. Breger, P. Gurfil, K.T. Alfriend, S.R. Vadali, J.P. How, *Spacecraft Formation Flying* (Elsevier Ltd, New York, 2010)
6. G. Gómez, J.M. Mondelo, The dynamics around the collinear equilibrium points of the RTBP. Phys. D: Nonlinear Phenom. **157**(4), 283–321 (2001)
7. G. Gómez, J. Llibre, R. Martínez, C. Simó, *Dynamics and Mission Design Near Libration Points*, Vol. I Fundamentals: The Case of Collinear Libration Points (World Scientific, Singapore, 2001)
8. G. Gómez, M. Marcote, High-order analytical solutions of Hill's equations. Celest. Mech. Dyn. Astron. **94**(2), 197–211 (2006)
9. À. Haro, M. Canadell, J.L. Figueras, A. Luque, J.M. Mondelo, *The Parameterization Method for Invariant Manifolds* (Springer, Berlin, 2016)
10. M.E. Hough, Lorentz force perturbations of a charged ballistic missile, in *Proceedings of the AIAA Guidance and Control Conference, AIAA-1982-1549, San Diego, CA* (1982)
11. R.S. Irving, *Integers, Polynomials, and Rings* (Springer, Berlin, 2004)
12. À. Jorba, J. Villanueva, On the persistence of lower Cdimensional invariant tori under quasiCperiodic perturbations. J. Nonlinear Sci. **7**, 427–473 (1997)
13. L.B. King, G.G. Parker, S. Deshmukh, J.H. Chong, Study of interspacecraft Coulomb forces and implications for formation flying. J. Propuls. Power **19**(3), 497–505 (2003)
14. E. Kolemen, N.J. Kasdin, P. Gurfil, Multiple Poincaré sections method for finding the quasiperiodic orbits of the restricted three body problem. Celest. Mech. Dyn. Astron. **112**(1), 47–74 (2012)
15. E. Kong, D. Kwon, S. Schweighart, L. Elias, R. Sedwick, D. Miller, Electromagnetic formation flight for multi-satellite arrays. J. Spacecr. Rocket. **41**(4), 659–666 (2004)
16. D.W. Kwon, Propellantless formation flight applications using electromagnetic satellite formations. Acta Astronaut. **67**(9–10), 1189–1201 (2010)
17. D.W. Kwon, R.J. Sedwick, Cryogenic heat pipe for cooling high-temperature superconducting coils. J. Thermophys. Heat Transf. **23**(4), 732–740 (2012)
18. K. Meyer, G. Hall, D. Offin, *Introduction to Hamiltonian Dynamical Systems and the N-Body Problem* (Springer Science & Business Media, Berlin, 2008)
19. M.A. Peck, Prospects and challenges for Lorentz-augmented orbits, in *Proceedings of the AIAA Guidance, Navigation, and Control Conference, AIAA-2005-5995, San Francisco, CA* (2005)
20. M.A. Peck, B. Streetman, C.M. Saaj, V. Lappas, Spacecraft formation flying using Lorentz forces. J. Br. Interplanet. Soc. **60**(7), 263–267 (2007)
21. C. Peng, *Relative motion and satellite formation based on Lorentz force (in chinese)* (Graduate University of Chinese Academy of Sciences, Master diss., 2012)
22. C. Peng, Relative orbital motion of a charged object near a spaceborne radially-directed rotating magnetic dipole, in*66th International Astronautical Congress, Jerusalem* (2015)
23. C. Peng, Y. Gao, Formation-flying planar periodic orbits in the presence of inter-satellite Lorentz force. IEEE Trans. Aerosp. Electron. Syst. (2017)
24. A.K. Porter, D.J. Alinger, R.J. Sedwick, J. Merk, R.A. Opperman, A. Buck, G. Eslinger, P. Fisher, D.W. Miller, E. Bou, Demonstration of electromagnetic formation flight and wireless power transfer. J. Spacecr. Rocket. **51**(6), 1914–1923 (2014)
25. Y. Ren, J.J. Masdemont, M. Marcote, G. Gómez, Computation of analytical solutions of the relative motion about a Keplerian elliptic orbit. Acta Astronaut.**81**(1), 186–199 (2012)

26. C.M. Saaj, V. Lappas, D. Richie, M. Peck, B. Streetman, H. Schaub, Electrostatic forces for satellite swarm navigation and reconfiguration - Final report, *Final Report for Ariadna Study Id. AO491905* (2006)
27. L. Schaffer, J.A. Burns, Charged dust in planetary magnetospheres: Hamiltonian dynamics and numerical simulations for highly charged grains. J. Geophys. Res. **9**(A9), 17211–17223 (1994)
28. C. Simó: On the analytical and numerical approximation of invariant manifolds. In D. Benest, C. Froeshlé, (eds.), *Modern methods in Celestial Mechanics*: 285C330. Editions Frontires (1990)
29. L.A. Sobiesiak, C.J. Damaren, Controllability of Lorentz-augmented spacecraft formations. J. Guid. Control. Dyn. **38**(11), 2188–2195 (2015)
30. J. Stoer, R. Bulirsch, *Introduction to Numerical Analysis*, vol. 12 (Springer Science & Business Media, Berlin, 2013)
31. B. Streetman, M.A. Peck, New synchronous orbits using the geomagnetic Lorentz force. J. Guid., Control. Dyn. **30**(6), 1677–1690 (2007)
32. B. Streetman, M.A. Peck, A general bang-bang control method for lorentz augmented orbits, in *AAS Spaceflight Mechanics Meeting, AAS Paper 08-111, Galveston, Texas* (2008)
33. B. Streetman, M.A. Peck, Gravity-assist maneuvers augmented by the Lorentz force. J. Guid. Control Dyn. **32**(5), 1639–1647 (2009)
34. S. Tsujii, M. Bando, H. Yamakawa, Spacecraft formation flying dynamics and control using the geomagnetic Lorentz force. J. Guid. Control Dyn. **36**(1), 136–148 (2012)
35. A. Umair, D.W. Miller, J.L. Ramirez, Control of electromagnetic satellite formations in near-Earth orbit. J. Guid. Control Dyn. **33**(6), 1883–1891 (2010)
36. D. Vokrouhlicky, The Geomagnetic effects on the motion of electrically charged artificial satellite. Celest. Mech. Dyn. Astron. **46**(1), 85–104 (1989)
37. Y. Yu, H. Baoyin, Y. Jiang, Constructing the natural families of periodic orbits near irregular bodies. Mon. Not. R. Astron. Soc. **453**(3), 3269–3277 (2015)

Chapter 5
1:1 Ground-Track Resonance in a 4th Degree and Order Gravitational Field

In recent years, space missions with the destination of small solar system bodies have become more and more important. For missions to such kind of bodies, one of the biggest challenges comes from the perturbation on the spacecraft's motion by the highly irregular gravitational field. Therefore the dynamical environment in the vicinity of these bodies needs to be characterized.

In this chapter, using a gravitational field truncated at the 4th degree and order, the 1:1 ground-track resonance is studied. To address the main properties of this 1:1 resonance, a 1-degree of freedom (1-DOF) system is firstly studied. It is completely integrable. Equilibrium points (EPs), stability and resonance width are obtained. Different from previous studies, the inclusion of non-spherical terms higher than degree and order 2 introduces new phenomena. For a further study about the 1:1 resonance, a 2-DOF system is introduced, which includes the 1-DOF system and a second resonance acted as a perturbation part. With the aid of Poincaré section, the generation of chaos in the phase space is studied in detail by addressing the overlap process of these two resonances with arbitrary combinations of eccentricity (e) and inclination (i). Retrograde orbits, near circular orbits and near polar orbits are found to have better stability against the perturbation of the second resonance. The situations of complete chaos are estimated in the $e - i$ plane. By applying the maximum Lyapunov Characteristic Exponent (LCE), chaos is characterized quantitatively and same conclusions can be achieved. This study is applied to three asteroids, i.e. 1996 HW1, Vesta and Betulia, but the conclusions are not restricted to them.

5.1 Introduction

The commensurability (usually a ratio of simple integers) between the rotation of the primary body and the orbital motion of the surrounding spacecraft or particle is called ground-track resonance. A large amount of research has been carried out about the

J. Yuan et al., *Low Energy Flight: Orbital Dynamics and Mission Trajectory Design*, https://doi.org/10.1007/978-981-13-6130-2_5

1:1 ground-track resonances, i.e., the geosynchronous orbits. For example, a 2-DOF Hamiltonian system was modeled [5] near the critical inclination perturbed by the inhomogeneous geopotential. Global dynamics were studied in terms of Poincaré maps in the plane of inclination and argument of pericenter. Chaotic motions were expected close to the separatrix of the resonance of the mean motion.

However, for ground-track resonances in the highly irregular gravitational field (mainly small solar system bodies), the studies are limited. Scheeres [15] studied the stability of the 1:1 mean motion resonance with a rotating asteroid using a triaxial ellipsoid model, and applied it to Vesta, Eros and Ida. Later on, he studied the effect of the resonance between the rotation rate of asteroid Castalia and the true anomaly rate of an orbiting particle at periapsis with a 2nd degree and order gravitational field [16]. This kind of resonance was proven to be responsible for significant changes of orbital energy and eccentricity, and provides a mechanism for an ejected particle to transfer into a hyperbolic orbit or vice versa. Considering the 2nd degree and order gravitational field, Hu [8] showed that orbital resonance plays a significant role in determining the stability of orbits. Further, by modelling the resonant dynamics in a uniformly rotating 2nd degree and order gravitational field as a 1-DOF pendulum Hamiltonian [13], widths of the resonance were obtained in analytical expressions and also tested against numerical simulations for five resonances. They were found to be independent of the rotation rate and mass of the central body but strongly dependent on e and i. The retrograde orbits have a smaller resonance region than the prograde ones. In a slowly rotating gravitational field, the orbital stability was explained by the distance between the resonances but not by the strength of a specific one using the overlap criteria.

The resonant structure is explained with the truncated model for the equatorial and circular cases, respectively. Delsate [4] built the 1-DOF Hamiltonian of the ground-track resonances of Dawn orbiting Vesta. The locations of the EPs and the resonance width were obtained for several main resonances (1:1, 1:2, 2:3 and 3:2). The results were checked against numerical tests. The 1:1 and 2:3 resonances were found to be the largest and strongest ones, respectively. The probability of capture in the 1:1 resonance and escape from it was found to rely on the resonant angle. Tzirti [18] extended Delsate's work by introducing C_{30} into the 1:1 resonance, which resulted in 2-DOF dynamics. The C_{30} term was found to create tiny chaotic layers around the separatrix but without significant influence on the resonance width. With the ellipsoid shape model [3], MEGNO (Mean Exponential Growth factor of Nearby Orbits) was applied as an indicator to detect stable resonant periodic orbits and also 1:1 and 2:1 resonance structures under different combinations of the three semi-axes of the ellipsoid. A 1-DOF resonant model parametrized by e and i was obtained with a truncated ellipsoidal potential up to degree and order 2.

For the previous studies, the limitations are either the gravitational field which is truncated at degree and order 2 or the orbit which is restricted to a circular or polar case. In this chapter, the harmonic coefficients up to degree and order 4 are taken into account for studying the 1:1 resonance at different combinations of e and i, which results in a 2-DOF model. Therefore, this chapter is arranged as follows. Firstly, a 1-DOF Hamiltonian is built to investigate the main properties of the 1:1

resonance. The location of EPs and their stability are solved numerically for different combinations of e and i for Vesta, 1996 HW1 and Betulia. The resonance widths of the stable EPs are found numerically. Secondly, a 2-DOF Hamiltonian is introduced with the inclusion of a second resonance, which is treated as a perturbation on the 1-DOF Hamiltonian. Chaos is generated due to the overlap of the two resonances. By applying Poincaré sections, for all three asteroids, the extent of chaotic region in the phase space is examined against the distance between the primary and second resonances and their respective strengths. The roles that e and i play on the evolution of chaos in the phase space are studied systemically. Finally, the maximal LCE (mLCE) of the orbits in the chaotic seas are calculated for a quantitative study.

5.2 Dynamical Model

5.2.1 Hamiltonian of the System

The gravity potential expressed in orbital elements $(a, e, i, \Omega, \omega, M)$ is given by [9] as

$$V = \frac{\mu}{r} + \sum_{n\geq 2}^{\infty}\sum_{m=0}^{n}\sum_{p=0}^{n}\sum_{q=-\infty}^{\infty} \frac{\mu R_e^n}{a^{n+1}} F_{nmp}(i) G_{npq}(e) S_{nmpq}(\omega, M, \Omega, \theta) \tag{5.1}$$

in which μ and R_e are the gravitational constant and reference radius of the body, respectively. r is the distance from the point of interest to the center of mass of the body. $F(i)$ and $G(e)$ are functions of inclination and eccentricity, respectively. The complete list of them up to degree and order 4 can be found in [9]. In addition, n, m, p, q are all integers, θ is the sidereal angle.

$$S_{nmpq} = \begin{bmatrix} C_{nm} \\ -S_{nm} \end{bmatrix}_{n-m(\text{odd})}^{n-m(\text{even})} \cos\Theta_{nmpq} + \begin{bmatrix} S_{nm} \\ C_{nm} \end{bmatrix}_{n-m(\text{odd})}^{n-m(\text{even})} \sin\Theta_{nmpq}$$

with Kaula's gravitational argument Θ_{nmpq} written as

$$\Theta_{nmpq} = (n-2p)\omega + (n-2p+q)M + m(\Omega - \theta)$$

Given the Delaunay variables

$$l = M, g = \omega, h = \Omega, L = \sqrt{\mu a}, G = L\sqrt{(1-e^2)}, H = G\cos(i)$$

the Hamiltonian of the system can be written as

$$\mathscr{H} = T - V + \dot{\theta}\Lambda \tag{5.2}$$

in which $T = -\mu^2/2L^2$ is the kinetic energy and $\dot{\theta}$ is the rotation rate of the asteroid and Λ is the momentum conjugated to θ. Resonances occur when the time derivative of $\dot{\Theta}_{nmpq} \approx 0$. The 1:1 resonance is studied in detail in the following sections.

5.2.2 1:1 Resonance

According to [9], to study the 1:1 resonance, the resonant angle is introduced and defined as $\sigma = \lambda - \theta$, with the mean longitude $\lambda = \omega + M + \Omega$. This resonance occurs at $\dot{\sigma} \approx 0$, which means that the revolution rate of the orbit is commensurate with the rotation rate of the asteroid. In addition, it should be noticed that the solution of this 1:1 resonance is actually represented by the equilibrium points (EPs) that are commonly studied in a rotating (or body-fixed) frame, and is the phase angle of the EPs in a rotating frame.

The spherical harmonics that contribute to this resonance are listed in Appendix A. To introduce the resonant angle σ in the Hamiltonian and also keep the new variables canonical, a symplectic transformation is applied [19]

$$d\sigma L' + d\theta' \Lambda' = d\lambda L + d\theta \Lambda$$

and a new set of canonical variables is obtained as

$$\sigma, L' = L, \theta' = \theta, \Lambda' = \Lambda + L$$

After averaging over the fast variable θ', the Hamiltonian for the 1:1 resonance truncated at the second order of e can be written as

$$\mathscr{H} = \mathscr{H}_0 + \mathscr{H}_1 + \mathscr{H}_2 + o(e^3)$$

where $\mathscr{H}_k$, $k = 1, 2$ is the kth order of e and $\mathscr{H}_0$ is expressed as

$$\begin{aligned}
\mathscr{H}_0 = & -\frac{\mu^2}{2L^2} + \dot{\theta}(\Lambda' - L) - \frac{\mu^4 R^2}{L^6}\left[C_{20}(1-e^2)^{-\frac{3}{2}}\left(-\frac{1}{2} + \frac{3s^2}{4}\right)\right. \\
& \left. + \frac{3}{4}\left(1 - \frac{5e^2}{2} + \frac{13e^4}{16}\right)(1+c)^2(C_{22}\cos 2\sigma + S_{22}\sin 2\sigma)\right] \\
& - \frac{\mu^5 R^3}{L^8}\left[\left(1 + 2e^2 + 239\frac{e^4}{64}\right)\left(-\frac{3}{4}(1+c) + \frac{15}{16}(1+3c)s^2\right)(C_{31}\cos\sigma + S_{31}\sin\sigma)\right. \\
& \left. + \frac{15}{8}\left(1 - 6e^2 + \frac{423e^4}{64}\right)(1+c)^3(C_{33}\cos 3\sigma + S_{33}\sin 3\sigma)\right] \\
& - \frac{\mu^6 R^4}{L^{10}}\left[C_{40}(1-e^2)^{-\frac{7}{2}}\left(\frac{3}{8} - \frac{15s^2}{8} + \frac{105s^4}{64}\right)\right.
\end{aligned}$$

$$+\left(1+5e^2+\frac{65e^4}{16}\right)\left(-\frac{15}{8}(1+c)^2+\frac{105}{8}(c+c^2)s^2\right)(C_{42}\cos 2\sigma+S_{42}\sin 2\sigma)$$
$$\left.\frac{105}{16}\left(1-11e^2+\frac{199e^4}{8}\right)(1+c)^4(C_{44}\cos 4\sigma+S_{44}\sin 4\sigma)\right] \tag{5.3}$$

in which $c = \cos(i)$, $s = \sin(i)$ and L is used hereafter instead of L' for convenience. In terms of angular variables, it can be seen that $\mathcal{H}_0$ is only dependent on the angle σ. Since θ' is implicit in $\mathcal{H}_0$, its conjugate Λ' is a constant and can be dropped. Similarly, G and H, which are related to e and i, are constant as g, h are absent in $\mathcal{H}_0$. Therefore at a given combination of e and i, $\mathcal{H}_0$ is actually a 1-DOF system. However, $\mathcal{H}_1$ and $\mathcal{H}_2$ are functions of both σ and g and are 2-DOF systems. Their expressions are given in Appendix A and they are both zero at $e = 0$ or $i = 0$.

According to our simulations, it is found that $\mathcal{H}_1 \in o(\varepsilon^{3/2})$, $\mathcal{H}_2 \in o(\varepsilon)$, where ε is the ordering parameter and ranges from 10^{-2} to 10^{-1}. Since the origin of our selected body-fixed frame is also the center mass of the asteroid, the C_{21} and S_{21} terms are both zero, and the magnitude of $\mathcal{H}_1$ is therefore smaller than $\mathcal{H}_2$. Therefore, $\mathcal{H}_0$ with resonant angle σ can be viewed as the primary resonance. $\mathcal{H}_1$ and $\mathcal{H}_2$ with angles $j\sigma + kg$, $(j = 1, 2, 3, k = 1, 2)$ are the secondary resonances, which are expected to give rise to chaos.

5.3 Primary Resonance

5.3.1 EPs and Resonance Width

Firstly, $\mathcal{H}_0$ is studied in detail. Its equilibria can be found by numerically solving

$$\dot{\sigma} = \frac{\partial \mathcal{H}_0}{\partial L} = 0, \dot{L} = -\frac{\partial \mathcal{H}_0}{\partial \sigma} = 0 \tag{5.4}$$

The linearized system is written as

$$\begin{bmatrix} d\ddot{\sigma} \\ d\ddot{L} \end{bmatrix} = \begin{bmatrix} \frac{\partial^2 \mathcal{H}_0}{\partial L \partial \sigma} & \frac{\partial^2 \mathcal{H}_0}{\partial L^2} \\ -\frac{\partial^2 \mathcal{H}_0}{\partial \sigma^2} & -\frac{\partial^2 \mathcal{H}_0}{\partial L \partial \sigma} \end{bmatrix} \begin{bmatrix} d\sigma \\ dL \end{bmatrix}$$

The EPs are obtained by solving

$$\begin{cases} d\ddot{\sigma} = 0 \\ d\ddot{L} = 0 \end{cases}$$

The linear stability of an EP can be determined from the Jacobian matrix evaluated at the EP. The resonant frequency can be approximated at a stable EP (σ_s, L_s) as $\sqrt{\frac{\partial^2 \mathscr{H}_0}{\partial L^2}\frac{\partial^2 \mathscr{H}_0}{\partial \sigma^2}}|_{\sigma_s, L_s}$. Taking the Hamiltonian value corresponding to an unstable EP (σ_u, L_u), denoted as $\mathscr{H}_u$, its level curve on the phase map is actually the separatrix that divides the motion into libration and circulation regions [12]). Along this curve, L passes through its maximum L_{max} and also minimum L_{min} at $\sigma = \sigma_s$. The resonance width is then calculated as $\Delta L = L_{max} - L_{min}$ and is therefore only valid for the stable EPs.

5.3.2 Numerical Results

In this section, the EPs, their stability and the resonance width of asteroids Vesta, 1996 HW1 and Betulia are studied. They are selected because the first two asteroids are representatives of regular and highly bifurcated bodies, respectively, while Betulia has a triangular shape leading to large 3rd degree and order harmonics. The 4th degree and order spherical harmonics of the three asteroids are given in Appendix B. It is noted here that all the angles in this study are in radian. First, the dynamics due to the 2nd degree and order harmonics (C_{20} and C_{22}) is studied, hereafter denoted as $\mathscr{H}_{0/2nd}$.

(a) Vesta

As already mentioned in Sect. 5.2.2, the 1:1 resonance is actually the EPs in the rotating frame. Figure 5.1 gives the mean semi-major axis of the 1:1 resonance in the $e - i$ plane. The EPs with $\sigma = 0$ are unstable, while the ones with $\sigma = \pi/2$ are stable. The location of the EPs are symmetric with respect to $i = \pi/2$, due to the symmetry property of the 2nd degree and order gravity field. It is closest to Vesta when the orbit is polar and then gradually moves further away when the orbit approaches the equatorial plane (either prograde or retrograde). The resonance width decreases with the increase of i, and finally becomes zero when i arrives at π. This can be explained by the coefficient of the resonant angle 2σ in Eq. (5.3), denoted as f_{22}

$$f_{22} = -\frac{3\mu^4 R^2}{4L^6}\left(1 - \frac{5e^2}{2} + \frac{13e^4}{16}\right)(1+c)^2$$

When i approaches π, the term $1 + c$ becomes zero and f_{22} also comes to zero. For a given value of i, the larger e the smaller f_{22} is and the resonance width also decreases (Fig. 5.1). However, this phenomenon is weakened for larger i as its weight factor $(1 + c)^2$ becomes smaller. This can clearly be observed from the contour map. In addition, our results for orbits at $e = 0, i = 0$ or $e = 0, i = \pi/2$ are identical to those obtained in Delsate's study [4].

To investigate the effects of higher-order terms, the Hamiltonian $\mathscr{H}_0$ that includes the 4th degree and order harmonics is studied. Firstly, the phase portrait for some

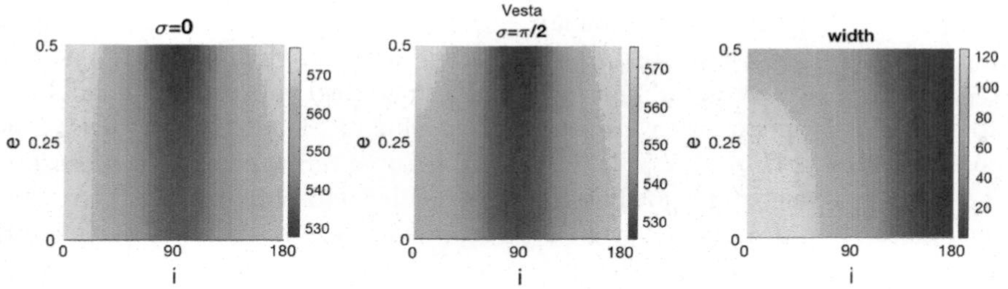

Fig. 5.1 The contour plots of mean semi-major axis (in km) of the unstable ($\sigma = 0$) and stable ($\sigma = \pi/2$) 1:1 resonance (the EPs) in the $e - i$ plane and the corresponding resonance width of stable EPs

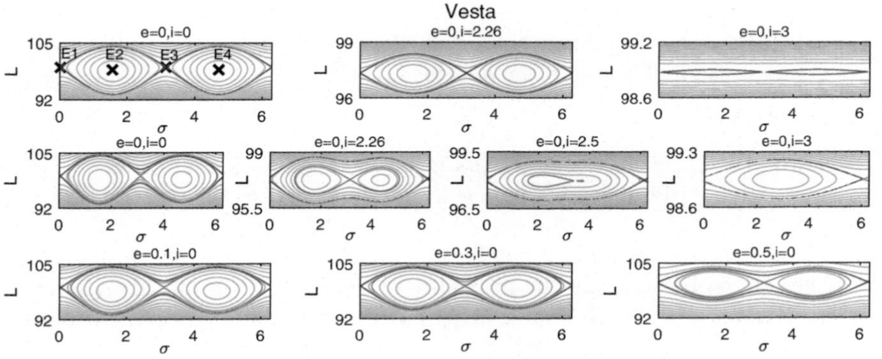

Fig. 5.2 The phase portrait of the Hamiltonian of Vesta. Top row: $\mathscr{H}_{0/2nd}$ for $e = 0, i = 0, 129.5°, 171.9°$; middle row: $\mathscr{H}_0$ for $e = 0, i = 0, 129.5°, 143.2°, 171.9°$; bottom row: $\mathscr{H}_0$ for $i = 0, e = 0.1, 0.3, 0.5$. The blue and red lines are the separatrix (or the values of the Hamiltonian) of the unstable EPs

example orbits with different e and i is given in Fig. 5.2. The four EPs are marked out as E1, E2, E3 and E4. The top three plots are actually the phase portrait of $\mathscr{H}_{0/2nd}$ for comparison, and the remaining ones are those of $\mathscr{H}_0$. It can be seen that due to the inclusion of 3rd and 4th harmonics, the symmetry with respect to $\sigma = \pi$ is broken and the EPs have a shift from $\sigma = 0$ and $\sigma = \pi/2$ but are still located in the near vicinity of them. Similar phenomena can be found for inclined geosynchronous orbits around Earth (see [14]). For the subplots in the middle, with the enlargement of i, the two stable EPs gradually merge into one and the unstable EP around $\sigma = \pi$ disappears, as a result of the increasing strength of harmonic coefficients other than C_{22}. The coefficient of the C_{31} term in Eq. (5.3) (denoted as f_{31}) is a first-order expression of $1 + c$ while for that of C_{22} it is of the second order $(1 + c)^2$.

$$f_{31} = -\frac{\mu^5 R^3}{L^8}\left(1 + 2e^2 + \frac{239e^4}{64}\right)\left(-\frac{3}{4}(1 + c) + \frac{15}{16}(1 + 3c)s^2\right)$$

When i is large enough, the influence of C_{31} on the structure of the phase space exceeds that of C_{22}. Therefore the phase space is dominated by the phase angle σ of C_{31}, and the existence of only one stable EP can be easily understood. In addition, all four EPs disappear when i approaches π, which is due to the fact that all the coefficients of $k\sigma(k = 1, 2, 3, 4)$ in Eq. (5.3) become zero when $i = \pi$ because of the terms $1 + c$ and s^2 and the phase portrait is filled with straight lines. The transit inclination (from four EPs to two EPs) is approximately 2.5 at $e = 0$ and slightly decreases to 2.2 at $e = 0.5$. This is explained by the fact that the C_{31} dynamics is strengthened when e becomes larger, which is witnessed by the fact that the large e promotes the merger of the two stable EPs shown in the bottom plots of Fig. 5.2. In addition, the larger e is, the smaller the value of the resonance width is, which can be explained by f_{22}. The resonance width for large i values is larger than that of the $\mathscr{H}_{0/2nd}$ dynamics, which can be explained by the dynamics taken over by C_{31} from C_{22} in these regions. The exact values of the EPs and the resonance width are given as a contour map in the $e - i$ plane in Appendix C.

(b) 1996 HW1

For 1996 HW1, the phase portrait of $\mathscr{H}_{0/2nd}$ and $\mathscr{H}_0$ is given in Fig. 5.3. There are four unstable EPs appearing in the equatorial plane ($i = 0$), which is consistent with the studies [6] and the results in [11]. They are also marked out as E1, E2, E3 and E4. For small i, there is no region for libration and therefore all the EPs are unstable. It can be seen that the instability of the four EPs is already determined by the dynamics of $\mathscr{H}_{0/2nd}$. The inclusion of other harmonics however has a strong distortion of the phase space. Two of the unstable EPs become stable at $i \approx 108.9°$ for $\mathscr{H}_{0/2nd}$ and at $i \approx 120.3°$ for $\mathscr{H}_0$, indicating the destabilizing effects of the highly irregular gravitational fields and the stability of the retrograde motion in this highly perturbed environment. Then the two EPs merge into one at $i \approx 154.7°$ also due to strong effects of C_{31} and finally disappear for the same reason: the effects of terms

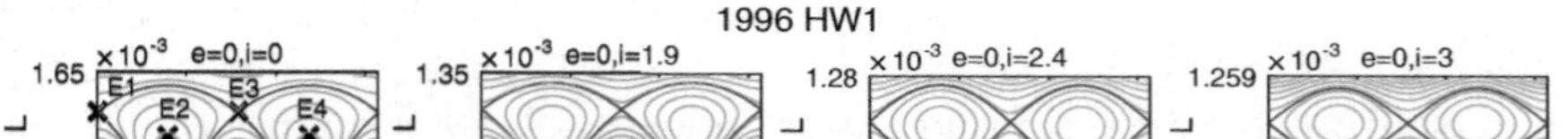

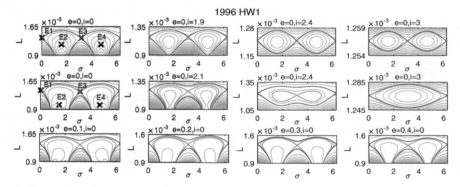

Fig. 5.3 The phase portrait of the Hamiltonian of 1996 HW1. Top row: $\mathscr{H}_{0/2nd}$ for $e = 0, i = 0, 108.9°, 137.5°, 171.9°$; middle row: $\mathscr{H}_0$ for $e = 0, i = 0, 120.3°, 137.5°, 171.9°$; bottom row: $\mathscr{H}_0$ for $i = 0, e = 0.1, 0.2, 0.3, 0.4$. The blue and red lines are the separatrix of the unstable EPs

$1 + c$ and s^2. The phase portrait is slightly influenced by e with the exception that the elongated orbit (with larger e) has been less influenced by the high degree and order harmonics, as shown in the bottom subplots.

The σ of the EPs and the resonance width are only obtained for the situation where stable EP exists and are given in Appendix C. The semi-major axis of the stable EPs and the unstable EPs are also given, indicated by a_s and a_u, respectively. After arriving at the maximum value at $i \approx 128.9°$, the resonance width decreases and becomes zero when i approaches π. However, it is not affected by e, as the dynamics is mainly dominated by i rather than e. Therefore the most interesting range for resonance is within $126° < i < 171.9°$, which will be further studied in the next section.

(c) Betulia

The phase portrait of Betulia is given in Fig. 5.4. Only four EPs (E1, E2, E3 and E4) appear for $\mathscr{H}_{0/2nd}$, while there are six EPs apparent in the equatorial plane for $\mathscr{H}_0$ due to the triangular shape of this body. Among them, E2, E4 and E6 are stable and E1, E3 and E5 are unstable. In Magri's study [10], six EPs were also obtained near the equatorial plane of Betulia, but using a polyhedron gravitational field. However, they found E6 unstable while it is stable (the right stretched island) from our phase portrait. The second difference is that their EPs are in general slightly closer to the body than ours. These distinctions primarily originate from the different gravitational fields applied in the studies. However, the gravitational field truncated at degree and order 4 applied in this study already captures the main dynamical properties of the complete gravitational field to a large extent at least for 1:1 resonant dynamics, which is the focus of our study.

As shown in the middle row of Fig. 5.4, the phase portrait of $\mathscr{H}_0$ changes significantly with the increase of i. There are actually three EPs within the left main island, which are illustrated as E2, E3 and E4 in the middle left plot. Among them, E2 and

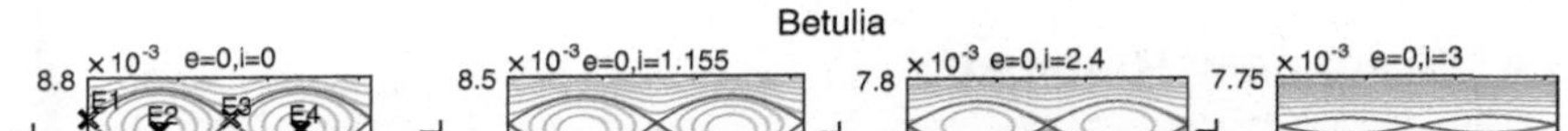

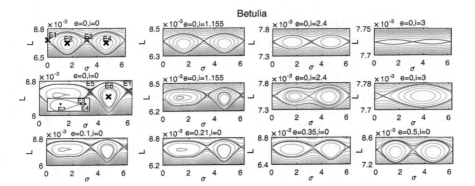

Fig. 5.4 The phase portrait of the Hamiltonian of Betulia. Top row: $\mathscr{H}_{0/2nd}$ for $e = 0, i = 0, 66.2°, 137.5°, 171.9°$; middle row: $\mathscr{H}_0$ for $e = 0, i = 0, 66.2°, 137.5°, 171.9°$; bottom row: $\mathscr{H}_0$ for $i = 0, e = 0.1, 0.21, 0.35, 0.5$. The blue and red lines are the separatrix of the unstable EPs

E4 are stable, while E3 is unstable and the most inner red line is its separatrix. From $i = 66.2°$, the unstable EP will disappear and the two stable ones start to merge, as can be seen clearly at $i = 137.5°$ with only two EPs left. Similarly, when i gets close to π, only one EP exists due to the dominant effect of C_{31}. In addition, because of the triangular shape of Betulia, the C_{31}, S_{31} and S_{33} terms are large compared to other asteroids, e.g. Vesta and 1996 HW1. Although S_{33} is one order of magnitude smaller than C_{22}, the coefficients of their phase angles 3σ and 2σ respectively are comparable with each other for small e and i. It is the S_{33} term that introduces two more EPs and also makes the phase space asymmetric with respect to $\sigma = \pi$. With the increase of both e and especially i, the influence of C_{31} becomes much stronger than that of C_{22} and S_{33} and finally dominates the phase space, which is the same as the cases of Vesta and 1996 HW1. In addition, the right island where EP6 is located is always slightly larger than the left one. In the next section, the resonance width of Betulia is actually measured as the width of the larger one. The exact values of σ of the EPs, the corresponding semi-major axes and the resonance width are also given in Appendix C.

In summary, this 1-DOF Hamiltonian $\mathscr{H}_0$ captures the main dynamics of the system and is illustrated with the three study cases above. The strength of C_{31} exceeds C_{22} and dominates the structure of the phase space when i approaches 180°. The large S_{33} term not only brings about more EPs but also introduces significant asymmetry of the phase space with respect to $\sigma = \pi$. In addition, due to the $1 + c$ term in the coefficients of all phase angles of $\mathscr{H}_0$, the resonance width of the retrograde orbits are found to be smaller than that of the prograde ones, which is consistent with Olsen's conclusion [13]. It was also found that the stability of the EPs is largely determined by the 2nd degree and order harmonics, especially for Vesta and 1996 HW1. However, the higher degree and order harmonics change the resonance width. The odd harmonics introduce new EPs to the system and break the symmetry of the 1:1 resonant dynamics.

5.4 Secondary Resonance

For a qualitative study about the effect of the second degree of freedom on the 1:1 resonant dynamics, $\mathscr{H}_1$ and $\mathscr{H}_2$ should be considered. However, the inclusion of all terms in $\mathscr{H}_1$ and $\mathscr{H}_2$ is far from trivial. For this study, the dominant term of $\mathscr{H}_2$ is taken into account. The dominant term, which has the largest amplitude, is given by

$$\mathscr{H}_{2d} = -\frac{\mu^4 R^2}{L^6} F_{221} G_{212} [C_{22} \cos(2\sigma - 2g) + S_{22} \sin(2\sigma - 2g)] \tag{5.5}$$

In the current study, thus only $\mathscr{H}_0$ and the dominant term $\mathscr{H}_{2d}$ are taken into account and the resulting 2-DOF Hamiltonian is written as

$$\mathscr{H}_{2dof} = \mathscr{H}_0 + \mathscr{H}_{2d}$$

A new resonant angle $2\sigma - 2g$ is introduced in the dynamics in addition to $k\sigma$ ($k = 1, 2, 3, 4$). A formal way to deal with this system is to treat $\mathscr{H}_{2d}$ as a small perturbation to the integrable system $\mathscr{H}_{2dof} = \mathscr{H}_0$ [7]. However, in our study, the perturbation of $\mathscr{H}_{2d}$ is not limited to small values, due to the large variations of e and i.

According to [2, 12], the dynamics of $\mathscr{H}_{2dof}$ can be studied by observing the overlap process of nearby resonances using Poincare maps. To a first approximation, each resonance is considered separately, namely only its own resonant angle is taken into account and the other one is neglected. The first resonance $\mathscr{H}_{reson1}$ is actually $\mathscr{H}_0$, and the second resonance $\mathscr{H}_{reson2}$ is defined as

$$\begin{aligned}\mathscr{H}_{reson2} &= -\frac{\mu^2}{2L^2} - \dot{\theta}L - \frac{\mu^4 R^2}{L^6} F_{201} G_{210} C_{20} + \mathscr{H}_{2d} \\ &= -\frac{\mu^2}{2L^2} - \dot{\theta}L - \frac{\mu^4 R^2}{L^6}(1-e^2)^{-\frac{3}{2}}\left(-\frac{1}{2} + \frac{3s^2}{4}\right) C_{20} \\ &\quad - \frac{\mu^4 R^2}{L^6}\left(\frac{9e^2}{4} + \frac{7e^4}{4}\right) s^2 [C_{22}\cos(2\sigma - 2g) + S_{22}\sin(2\sigma - 2g)]\end{aligned}$$

which only includes one resonant angle $2\sigma - 2g$. Their locations need to be solved first and then the Poincaré maps of the single-resonance dynamics are computed respectively on the same section in the vicinity of their location. If $\mathscr{H}_{2d}$ is small enough, the separatrix of $\mathscr{H}_{reson2}$ is further away from that of $\mathscr{H}_{reson1}$ and the two resonances are slightly influenced by each other. Tiny chaotic layers are probably generated around the separatrix. Otherwise, if $\mathscr{H}_{2d}$ is large, the separatrix of the two resonances intersect, their dynamical domains overlap, and each resonance is significantly affected by the other one. The chaotic layers extend to large-region chaos that dominates the phase space. Since $\mathscr{H}_{reson1}$ is the dominant dynamics of our 1:1 resonant model, the focus is to put on how $\mathscr{H}_{reson1}$ is influenced by $\mathscr{H}_{reson2}$, which can also be interpreted as how much the 1-DOF dynamics is affected by a perturbation.

5.4.1 The Location and Width of $\mathscr{H}_{reson2}$

The location and width of $\mathscr{H}_{reson1}$ have been obtained in Sect. 5.2.2. Since we want to apply Poincaré sections to study the dynamics, the section of the map is first defined here as $g = \pi/2$, $\dot{g} > 0$ in the $L - \sigma$ plane. Since $\mathscr{H}_{reson1}$ has 1-DOF, its Poincaré map is the same as its phase portrait in the phase space. The location of $\mathscr{H}_{reson2}$ on this section can be obtained by numerically solving

$$\begin{cases} 2\dot{\sigma} - 2\dot{g} = \dfrac{2\partial\mathscr{H}_{reson2}}{\partial L} - \dfrac{2\partial\mathscr{H}_{reson2}}{\partial G} = 0 \\ \mathscr{H}_{reson2}(\sigma_0, g_0, L^*, G^*) = \mathscr{H}_{seperatrix} \end{cases} \tag{5.6}$$

in which $\sigma_0 = g_0 = \pi/2$. $\mathscr{H}_{seperatrix}$ is the Hamiltonian value of the seperatrix of $\mathscr{H}_{reson1}$ which is also the energy constant of the section. L^* and G^* represent the variables that need to be solved. As $\mathscr{H}_{reson2}$ itself is a 2-DOF system, the pendulum model cannot be applied for approximating its resonance width. Therefore, based on the dynamical properties of the Poincaré map, a full numerical estimation is used. By integrating from the initial point $(\sigma_0, g_0, L^*, G^*)$ for moderate iterations, a curve is obtained which is either the upper or the lower part of the separatrix of $\mathscr{H}_{reson2}$ on the section. If it is the upper part, L_{max} is directly obtained by taking record of the maximum point of the curve. L_{min} is the minimum of the lower border obtained by integrating from the point $(\sigma_0, g_0, L^* - \delta L, G^*)$ with a displacement from $(\sigma_0, g_0, L^*, G^*)$ by δL depending on the dynamics studied and vice versa. The curves acquired are the separatrix of $\mathscr{H}_{reson2}$. Therefore, the width of $\mathscr{H}_{reson2}$ is approximated by $L_{max} - L_{min}$, which is already good enough for the current study.

Given that the maxima and minima of $\mathscr{H}_{reson1}$ and $\mathscr{H}_{reson2}$ are denoted as L_{max1}, L_{min1} and L_{max2}, L_{min2}, respectively, the relative locations of the two resonances can be characterized by $L_{min1} - L_{max2}$ and $L_{min1} - L_{min2}$. The former one, which is the distance between the lower borders of the two resonances, is positive if the two resonances are totally separated and becomes negative as the resonances start to overlap with each other. The latter one is actually the measurement of the extent of overlap of the two resonances. Its non-positive value indicates that one resonance is completely within the other one. For different combinations of e and i, the 2-DOF Hamiltonian $\mathscr{H}_{2dof}$ is studied for 1996 HW1, Vesta and Betulia.

5.4.2 1996 HW1

Since 1996 HW1 only has a limited inclination range $126° \lesssim i < 180°$ of libration motion of $\mathscr{H}_{reson1}$ (shown in Fig. 5.3), its second degree of freedom dynamics is studied first. In Fig. 5.5, the upper plots give the separatrices of the two resonances on the Poincaré maps, which are the boundaries of their phase space. The bottom plots are the phase space of $\mathscr{H}_{2dof}$ on the same section, both for i changing from 171.9° to 143.2° at the example $e = 0.1$. Figure 5.5 reflects the relationship of the distance between the two resonances $\mathscr{H}_{reson1}$ and $\mathscr{H}_{reson2}$ and the extent of chaotic region of $\mathscr{H}_{2dof}$.

The Effect of *i*

For $i = 171.9°$, even though the resonances do not overlap (but are close), tiny chaotic layers appear in the vicinity of the separatrix of $\mathscr{H}_{2dof}$. When there is a small overlap at $i = 166.2°$, the chaotic layer is extended but a large libration region still retains. With the increase of the overlap from $i = 166.2°$, a large part of the phase space is occupied by chaos. The regular region shrinks to a limited area at the center of the phase space and meanwhile five islands appear around it, which is due to the high-order resonances between H_{reson1} and $\mathscr{H}_{reson2}$. With the further decrease of i to 158.7°, $\mathscr{H}_{reson2}$ is almost completely inside $\mathscr{H}_{reson1}$ and there are only three small

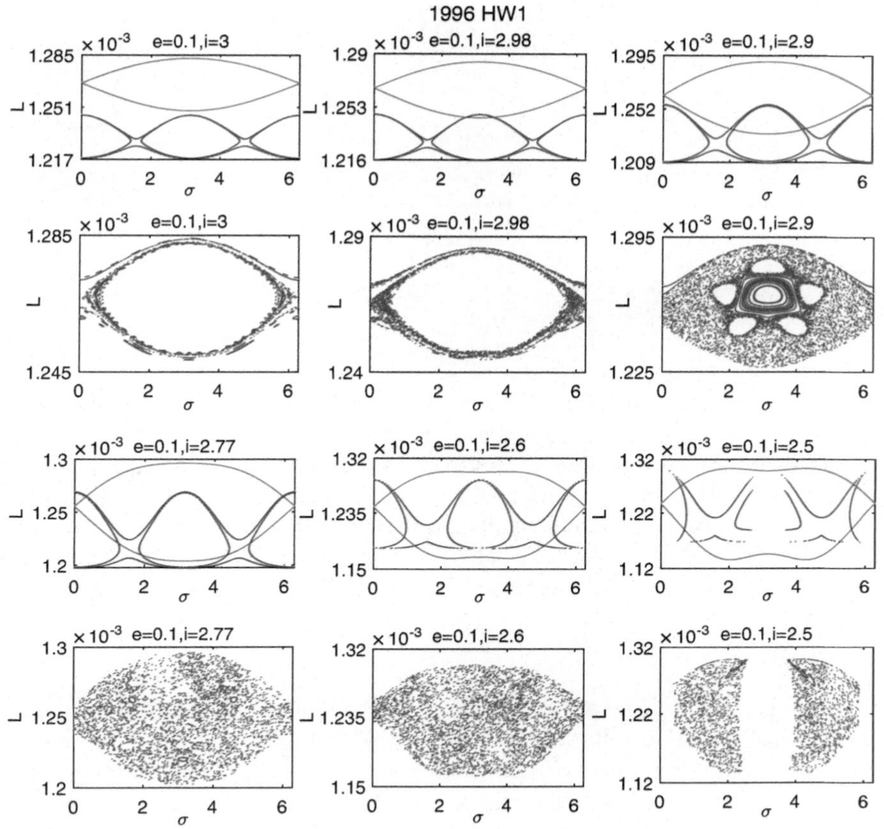

Fig. 5.5 First and third rows: the separatrices of resonances $\mathscr{H}_{reson1}$ (red) and $\mathscr{H}_{reson2}$ (blue) on the section $g = \pi/2$, $\dot{g} > 0$; second and fourth rows: the phase space of the corresponding $\mathscr{H}_{0/2dof}$; both for $e = 0.1$, $i = 171.9°, 170.7°, 166.2°, 158.7°, 149°, 143.2°$

KAM tori left, indicating the system is transiting to global chaos. In addition, the original stable EP becomes unstable as the center part is already chaotic. Although the dynamics is completely chaotic at $i = 149°$, the chaos is still bounded by the separatrix of H_{reson1}. However, finally at $i = 143.2°$ the whole structure of $\mathscr{H}_{reson1}$ could not be kept and the continuity of phase space is broken. It is noticed that this break is consistent with the break of the separatrix of $\mathscr{H}_{reson2}$ at the same range of σ, implying a significant perturbation of $\mathscr{H}_{reson2}$ on the total dynamics. The break of $\mathscr{H}_{reson2}$'s separatrix attributes to the fact that the time derivative of g (namely $\dot{g}$) changes its sign from positive to negative after i crossing some specific value, and therefore it produces no crossings on the section which is defined as $\dot{g} > 0$. This phenomenon will be discussed in detail in the next section.

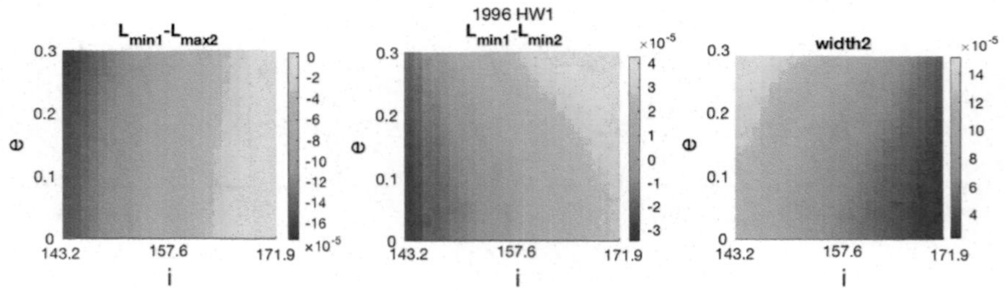

Fig. 5.6 The distance between $\mathscr{H}_{reson1}$ and $\mathscr{H}_{reson2}$ measured as $\mathscr{L}_{min1} - \mathscr{L}_{max2}$ (left) and $\mathscr{L}_{min1} - \mathscr{L}_{min2}$, and the width of $\mathscr{H}_{reson2}$ (right)

In summary, i has a great influence on the 2-DOF dynamics at a constant e. When i decreases, $\mathscr{H}_{reson2}$ is strengthened as it includes the term s^2 (as seen in Eq. (5.4)) and its resonance width increases. However, its location does not deviate much. For $\mathscr{H}_{reson1}$, not only its width is increasing but also its location is moving downward. Ultimately, the two resonances totally overlap and have a strong interaction with each other. Nevertheless, the width of $\mathscr{H}_{2dof}$ is determined by $\mathscr{H}_{reson1}$, which is seen from both L values, although the inner structure of the phase space has been totally affected.

The Effect of *e*

To study the effect of e on the dynamics, the contour map of the distance of the two resonances and also the width of the second resonance are given on the $e - i$ plane in Fig. 5.6. In the left plot, the yellow region indicates the situation of non-overlap and slight overlap. In the middle plot, the green and blue areas demonstrate the situation when $\mathscr{H}_{reson2}$ moves totally inside $\mathscr{H}_{reson1}$ and the overlap between the two is complete. The right plot demonstrates that the width of $\mathscr{H}_{reson2}$ is also enlarged when e becomes large, which can be proven by the term $9e^2/4 + 7e^4/4$ in $\mathscr{H}_{reson2}$. Therefore, the largest distance of $\mathscr{H}_{reson1}$ and $\mathscr{H}_{reson2}$ is witnessed at the down-right corner and $\mathscr{H}_{reson2}$ approaches its highest location at the upper-left corner in the left plot. In addition, as e increases and i decreases, $\mathscr{H}_{reson2}$ becomes stronger (as indicated by the resonance width) and has a significant influence on the dynamics of $\mathscr{H}_{reson1}$.

Therefore, given a specific e and i, an estimation from this contour map can be made on when small chaotic layers appear and when large chaotic seas are expected. As an example, for $e = 0.1$, tiny chaotic layers are apparent at $i = 171.9°$ when the two resonances start to overlap; the last KAM tori disappear and the phase space is full with chaos around $i = 158.2°$. For a more complete understanding, the phase space of $\mathscr{H}_{2dof}$ at $e = 0.3$ with different i is given in Fig. 5.7. As compared to Fig. 5.5, the upper plots of Fig. 5.7 show that the large e distorts the main island, which originally has circular or ellipsoidal shape. The chaos is more abundant and the size of the main island reduces and a new phase structure is generated at the bottom of the plot, due to the stronger influence of $\mathscr{H}_{reson2}$. In addition, the lower half of the chaos is thicker

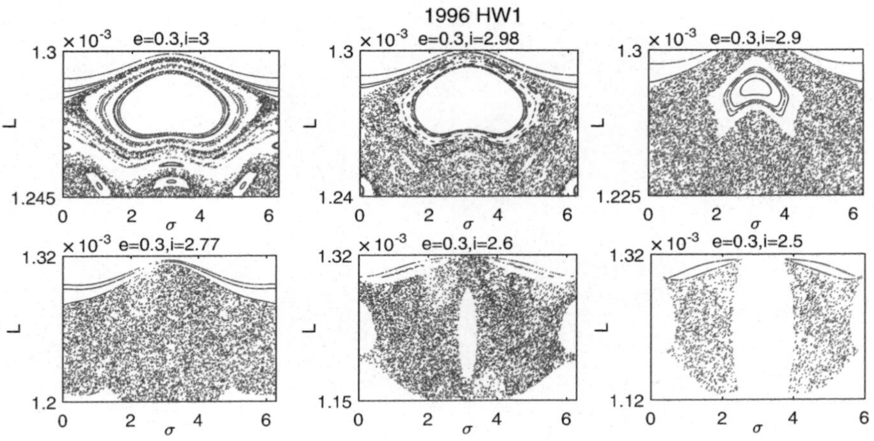

Fig. 5.7 The phase space of $\mathscr{H}_{2dof}$ at $e = 0.3$ for $i = 171.9°, 170.7°, 166.2°, 158.7°, 149°, 143.2°$

than the upper part, as it is more influenced by the perturbation from $\mathscr{H}_{reson2}$ which approaches $\mathscr{H}_{reson1}$ from the bottom direction. In addition, the islands appearing at the bottom area of the phase space can be explained by the direct interaction of $\mathscr{H}_{reson1}$ and $\mathscr{H}_{reson2}$ in that region. Furthermore, the lower three plots are full of chaos.

For $e = 0.3$, $i = 149°$, $\mathscr{H}_{reson2}$ is already comparable with $\mathscr{H}_{reson1}$ on the dynamics of $\mathscr{H}_{2dof}$. Therefore, the center part of the phase space is not completely chaotic; limited regular (white) region appears which is actually part of the center island of $\mathscr{H}_{reson2}$.

Bounded Chaotic Regions

As mentioned in Introduction, tiny chaotic layers are generated in the vicinity of the separatrix. Their boundaries can be estimated, which is the topic of this section.

For small perturbations, it is known from [12] that the chaotic region covers areas spanned by the instantaneous separatrices for varying secular angles, which is g in our study. Its boundaries are estimated from the separatrices corresponding to the minimal and maximal resonant width. This is also known as the modulated-pendulum approximation. The small perturbation corresponds to the case of close approach and almost contact between $\mathscr{H}_{reson1}$ and $\mathscr{H}_{reson2}$, and is therefore applicable to a situation with quite large inclination values. For 1996 HW1, Fig. 5.8 illustrates this region at different eccentricities and inclinations.

In each plot, the outer red lines represent the boundary corresponding to the Hamiltonian value of $\mathscr{H}_{seperatrix} + \mathscr{H}_{2d}(g = 0)$ which is the maximum resonance width. The inner red line is the inner boundary with the Hamiltonian value of $\mathscr{H}_{seperatrix} + \mathscr{H}_{2d}(g = \pi)$ which is the minimum resonance width. For $e = 0.1$, $i = 171.9°$, in which case the perturbation from $\mathscr{H}_{reson2}$ is the weakest (shown in Fig. 5.5), the theory works perfectly since all chaos is restricted to the region between the two

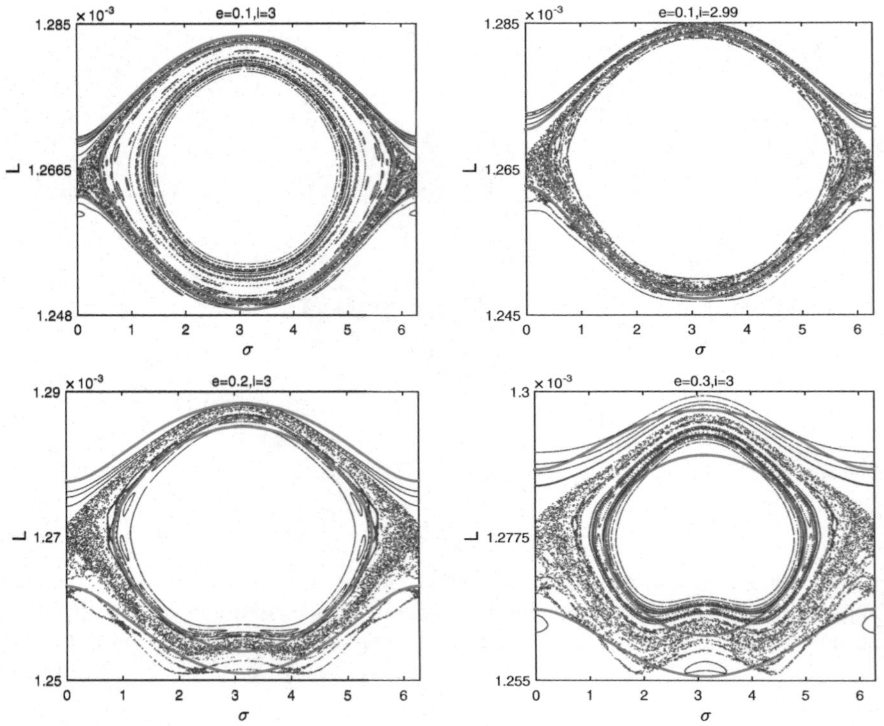

Fig. 5.8 The chaotic regions bounded by the separatrices (red lines) corresponding to the minimal and maximal resonant width for $e = 0.1, i = 171.9°, 171.3°; i = 171.9°, e = 0.2, 0.3$

red lines. When the inclination decreases to 171.3°, the chaos is still well bounded but a small portion of it at the bottom area is already outside the red lines. For comparison, the cases of $e = 0.2, i = 171.9°$ and $e = 0.3, i = 171.9°$ are studied, in which situation the perturbations of $\mathscr{H}_{reson2}$ are not small anymore. The chaotic region does not fit within the red lines well. The bottom part of the chaos is shifted upwards and is therefore outside the inner red line, due to the distortion of the phase space.

In summary, the chaotic layers are well estimated with the approximation theory for the small perturbation cases (i close to 180° and small e). The strong perturbation of $\mathscr{H}_{reson2}$ introduced by large e not only broadens the chaotic region, but also reshapes the phase space.

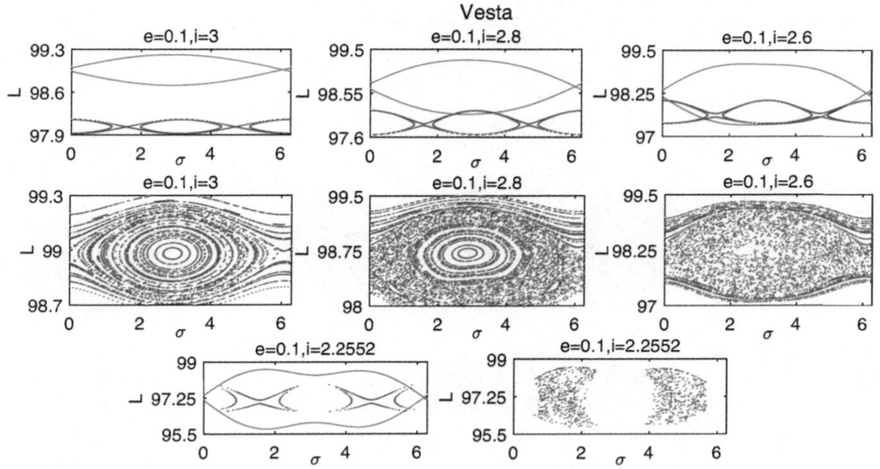

Fig. 5.9 First row: the separatrices of resonances $\mathscr{H}_{reson1}$ (red) and $\mathscr{H}_{reson2}$ (blue) on the section $g = \pi/2, \dot{g} > 0$; second row: the phase space of the corresponding $\mathscr{H}_{0/2dof}$; both for $e = 0.1$, $i = 171.9°, 160.4°, 149°$; third row: the separatrices and phase space for $e = 0.1, i = 129.2°$ (retrograde)

5.4.3 Vesta

The Effect of i

For Vesta, there is always a stable EP for different i and e. For the retrograde case, the distance between $\mathscr{H}_{reson1}$ and $\mathscr{H}_{reson2}$ and the phase space of $\mathscr{H}_{2dof}$ for orbits with different i but the same $e = 0.1$ is given in Fig. 5.9. For $i = 171.9°$, the dynamics of $\mathscr{H}_{reson1}$ is hardly influenced since the two resonances are far apart and there is no interaction between them. With the decreasing of i, a significant chaotic region around the separatrix appears, even when the two resonances are just in contact (as seen at $i = 160.4°$). When $\mathscr{H}_{reson2}$ completely evolves inside $\mathscr{H}_{reson1}$, the phase space becomes totally chaotic, as indicated at $i = 149°$. Further at $i = 129.2°$, the phase space becomes discontinuous and only scattered points are left without any recognizable dynamical structure (similar to the case for 1996 HW1 at $i = 143.2°$). The reason will be explained later this section.

The situation is quite different for the prograde case, as shown in Fig. 5.10. It can be seen that $\mathscr{H}_{reson2}$ is completely inside $\mathscr{H}_{reson1}$ for all inclinations; also the width of $\mathscr{H}_{reson2}$ increases as the orbit gets more inclined. When the strength of $\mathscr{H}_{reson2}$ is very weak at $i = 11.5°$, very tiny chaotic layer is present around the separatrix. When $\mathscr{H}_{reson2}$ becomes stronger at $i = 34.4°$, new islands inside the two main libration regions are generated, in addition to the weak chaos. Finally, when $\mathscr{H}_{reson2}$ is large enough at $i = 45.8°$, the original phase structure is broken and the two main islands are filled with chaos but not connected anymore. Combined with the previous analysis, the extent of chaos is found to be not only related to the location

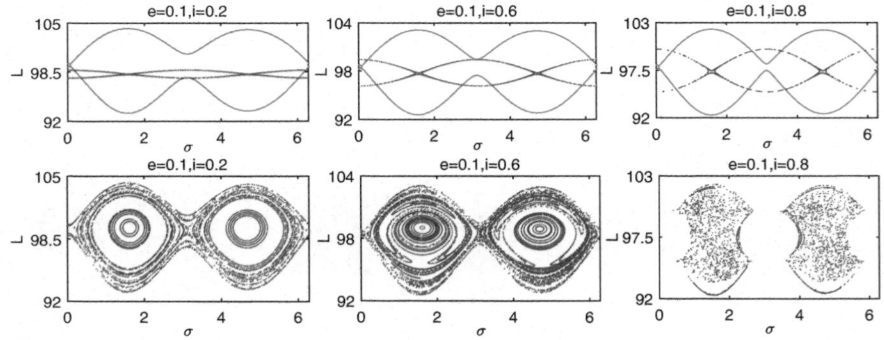

Fig. 5.10 First row: the separatrices of resonances $\mathscr{H}_{reson1}$ (red) and $\mathscr{H}_{reson2}$ (blue) on the section $g = \pi/2$, $\dot{g} > 0$; second row: the phase space of the corresponding $\mathscr{H}_{0/2dof}$; both for $e = 0.1$, $i = 11.5°, 34.4°, 45.8°$ (prograde)

of the two resonances, but their relative strength is also important. The dynamics of $\mathscr{H}_{2dof}$ is determined by the evolution (location, stability and strength) of both $\mathscr{H}_{reson1}$ and $\mathscr{H}_{reson2}$ as well as their interaction.

The Effect of *e*

Figure 5.11 shows contour maps that can be used to analyze the influence of e. For the retrograde case, the effects of e and i on the evolution of the two resonances are similar to that of 1996 HW1, as shown in the bottom plots of Fig. 5.11. The slight difference is that the maximum resonance width of $\mathscr{H}_{reson2}$ is at $e = 0.5$ and $i \approx 137.5°$ for Vesta rather than at the top-left corner for 1996 HW1, which can be explained by the non-linear property of the resonance width as a function of e and i. For the prograde case, instead of $L_{min1} - L_{max2}$, $L_{max1} - L_{max2}$ is obtained due to the fact that the two resonances already completely overlap. It is always positive for $L_{max1} - L_{max2}$ and negative for $L_{min1} - L_{min2}$. It must be mentioned that when $i = 0$, the width of $\mathscr{H}_{reson2}$ is zero and the dynamics of $\mathscr{H}_{reson1}$ is not affected. Therefore, we start our calculation from $i = 0.6°$. The largest distances between the maximum and minimum boundaries of the two resonances are both at the left-bottom corner of the contour map, and the smallest distances between them at the right-top corner, which can be easily explained by the corresponding weakest and strongest perturbing effect of $\mathscr{H}_{reson2}$. In addition, the width of $\mathscr{H}_{reson2}$ achieves its largest value at the largest inclination but smallest eccentricity. It can be noticed that the ranges of i stop at 51.6° and 128.9° for the prograde and retrograde orbits, respectively, due to the break of the separatrix of $\mathscr{H}_{reson2}$ at $51.6° \lesssim i \lesssim 128.9°$.

For a complete understanding, the phase space of $\mathscr{H}_{2dof}$ at $e = 0.5$ is given in Fig. 5.12. Similarly, compared to the phase space at $e = 0.1$ (shown in Figs. 5.9 and 5.10), the main island is strongly distorted and the chaotic region is significantly extended, due to the strong perturbation of $\mathscr{H}_{reson2}$. For $e = 0.5$, $i = 149°$, the regular region at the center of the phase space again is actually part of the regular region of

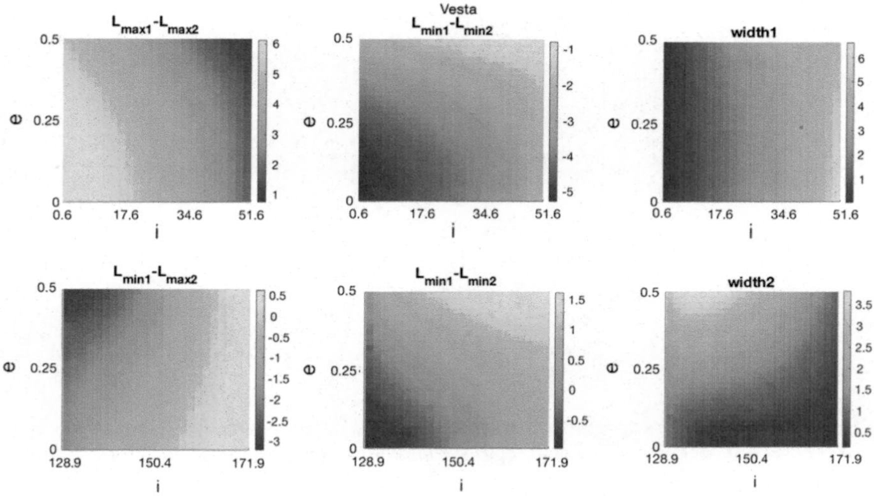

Fig. 5.11 The distance between $\mathscr{H}_{reson1}$ and $\mathscr{H}_{reson2}$ measured as $\mathscr{L}_{min1} - \mathscr{L}_{max2}$ (left) and $\mathscr{L}_{min1} - \mathscr{L}_{min2}$ (middle), and the width of $\mathscr{H}_{reson2}$ (right) for the prograde (top) and retrograde (bottom) cases (i in radian)

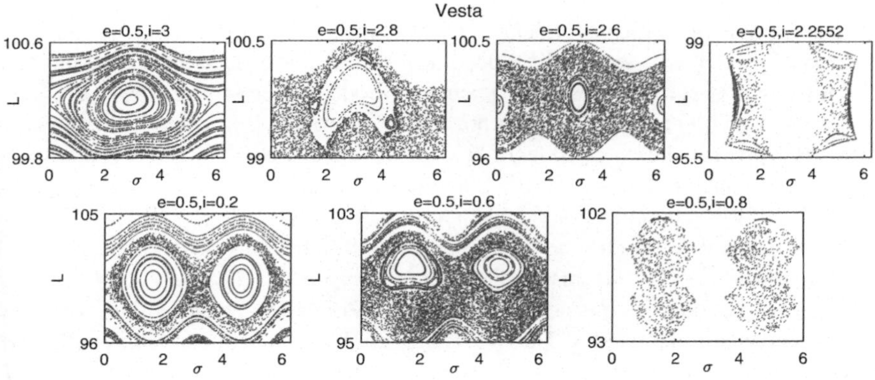

Fig. 5.12 The phase space of $\mathscr{H}_{2dof}$ at $e = 0.5$ for $i = 171.9°, 160.4°, 149°, 129.2°, 11.5°, 34.4°, 45.8°$

$\mathscr{H}_{reson2}$, due to the comparable influence of $\mathscr{H}_{reson1}$ and $\mathscr{H}_{reson2}$ on the dynamics of $\mathscr{H}_{2dof}$. Therefore, large e gives rise to strong perturbations on the dynamics.

Near Polar Region

For the near polar region, the dynamical structure shrinks and almost disappears on our previously defined section, as can already be seen from the plots at $i = 45.8°$ and $i = 129.2°$ in Figs. 5.9 and 5.10. This is due to the fact that the secular rate of g (Kaula 1966)

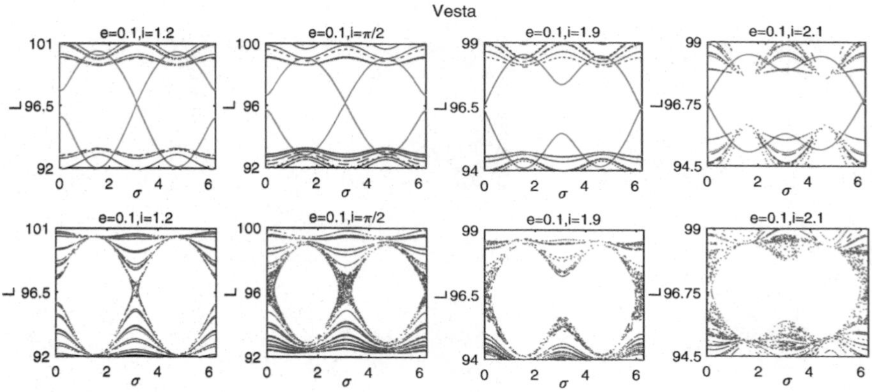

Fig. 5.13 First row: the separatrices of resonances $\mathscr{H}_{reson1}$ (red) and $\mathscr{H}_{reson2}$ (blue) on the section $g = \pi/2, \dot{g} > 0$; second row: the phase space of the corresponding $\mathscr{H}_{0/2dof}$; both for $e = 0.1, i = 68.8°, 90°, 108.9°, 120.3°$

$$\dot{g} = -\frac{3nR^2C_{20}}{a^2(1-e^2)^2}(4-5s^2) \tag{5.7}$$

changes sign at the critical inclinations $i_{critical} = 63.4°$ and $116.6°$. In particular, it is negative for $63.4° < i < 116.6°$. Since this formula of $\dot{g}$ is obtained from averaging the leading C_{20} perturbation and our current model includes additional harmonics terms, the sign of $\dot{g}$ in our study does not change sharply at $i_{critical}$ but has a transition process. However, the exact values of this transition are beyond the scope of this study. Some orbits may still have $\dot{g} > 0$ while others already have $\dot{g} < 0$, which explains the break of the separatrix of $\mathscr{H}_{reson2}$ on the section $g = \pi/2, \dot{g} > 0$. Therefore, we can define a new section for the near polar orbits with the only difference that $\dot{g} < 0$.

For similar simulations, results are shown in Fig. 5.13. However, no (L^*, G^*) of $\mathscr{H}_{reson2}$ can be found on this new section; and its Poincaré map rather than separatrix is included in plots labelled A. The two resonances have moderate overlap at the upper and lower boundaries of $\mathscr{H}_{reson1}$, which brings about limited chaotic regions closely attached to the separatrix. When $\mathscr{H}_{reson2}$ reaches its strongest effect at $i = 90°$, the chaos becomes more obvious and thick. For $i = 68.8°$ and $108.9°$, the chaos is visible but less abundant. For $i = 120.3°$, in addition to the chaos, islands are apparent in the circulation region of $\mathscr{H}_{reson1}$, where the two resonances have a strong modulation with each other. In conclusion, the libration region of $\mathscr{H}_{reson1}$ is hardly influenced by $\mathscr{H}_{reson2}$. From this point of view, the 1:1 resonance is more stable for near polar orbits at different eccentricities.

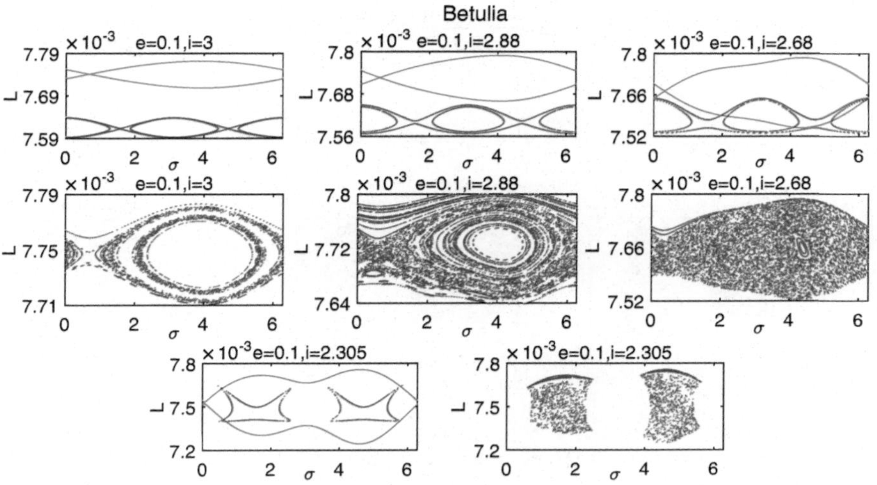

Fig. 5.14 First row: the separatrices of resonances $\mathcal{H}_{reson1}$ (red) and $\mathcal{H}_{reson2}$ (blue) on the section $g = \pi/2, \dot{g} > 0$; second row: the phase space of the corresponding $\mathcal{H}_{0/2dof}$; both for $e = 0.1, i = 171.9°, 165°, 153.6°$; third row: the separatrices and phase space for $e = 0.1, i = 132°$ (retrograde)

5.4.4 Betulia

For both prograde and retrograde orbits, Betulia has very similar properties as Vesta, concerning the distance between the two resonances and the width of $\mathcal{H}_{reson2}$.

The Effect of i

The retrograde case is illustrated in Fig. 5.14. The most obvious difference w.r.t. Fig. 5.9 is that the phase space is not symmetric with respect to $i = 90°$ anymore. When the two resonances are further apart at $i = 171.9°$, $\mathcal{H}_{reson1}$ is hardly influenced. When they are almost in contact with each other at $i = 165°$, thick chaotic layers are present together with small islands in the phase space. Before complete overlap, there is a small KAM-tori left at $i = 153.6°$ and the center region of the phase space is also distorted. After that at $i = 132°$, the phase space is totally chaotic and finally broken. For the prograde case, the two resonances always totally overlap, as shown in Fig. 5.15. When $\mathcal{H}_{reson2}$ is tiny and weak at $i = 11.5°$, the main structure of $\mathcal{H}_{reson1}$ is kept, with the difference that small chaotic regions appear near the separatrix and again new islands are generated inside the right main island. As $\mathcal{H}_{reson2}$ becomes stronger at $i = 34.4°$, the overlapping part of $\mathcal{H}_{reson1}$ is completely chaotic. Similarly, for $i = 45.8°$, the phase space of $\mathcal{H}_{reson1}$ is significantly broken and is left with a large gap, even without complete break of the separatrix of $\mathcal{H}_{reson2}$. This indicates the strong perturbation of $\mathcal{H}_{reson2}$ on the dynamics and the highly non-linear property of $\mathcal{H}_{2dof}$.

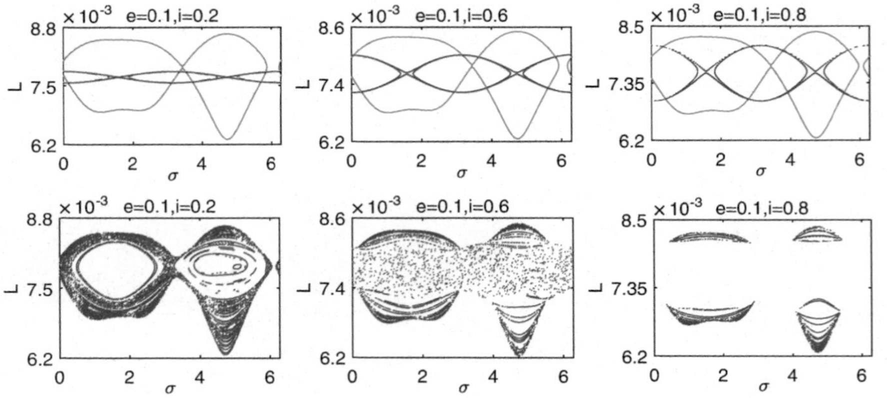

Fig. 5.15 First row: the separatrices of resonances $\mathscr{H}_{reson1}$ (red) and $\mathscr{H}_{reson2}$ (blue) on the section $g = \pi/2$, $\dot{g} > 0$; second row: the phase space of the corresponding $\mathscr{H}_{0/2dof}$; both for $e = 0.1$, $i = 11.5°, 34.4°, 45.8°$ (prograde)

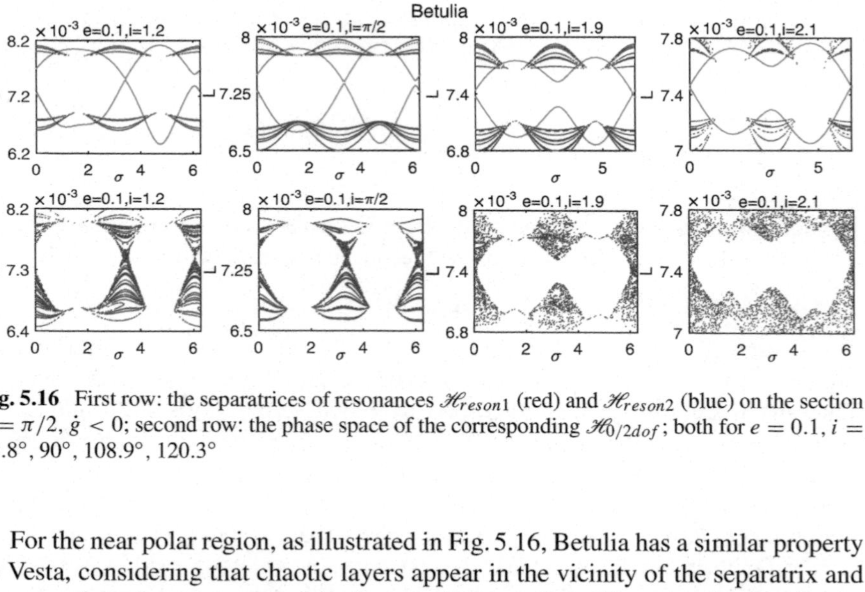

Fig. 5.16 First row: the separatrices of resonances $\mathscr{H}_{reson1}$ (red) and $\mathscr{H}_{reson2}$ (blue) on the section $g = \pi/2$, $\dot{g} < 0$; second row: the phase space of the corresponding $\mathscr{H}_{0/2dof}$; both for $e = 0.1$, $i = 68.8°, 90°, 108.9°, 120.3°$

For the near polar region, as illustrated in Fig. 5.16, Betulia has a similar property as Vesta, considering that chaotic layers appear in the vicinity of the separatrix and also new islands are generated in the circulation region. However, for Betulia the three regions are weakly connected at $i = 68.8°$. Furthermore, they become totally isolated at $i = 90°$, due to the stronger modulation of $\mathscr{H}_{reson2}$ compared to that of Vesta (Fig. 5.13). Since the regular region is open, the originally stable EPs of $\mathscr{H}_{reson1}$ probably change into unstable. At $i = 108.9°$ and $120.3°$, the circulation region is full of chaos, which implies that the perturbation of $\mathscr{H}_{reson2}$ and its interaction with $\mathscr{H}_{reson1}$ in this region are stronger, compared to the cases at $i = 68.8°$ and $90°$. Again, the general structure of the libration part of $\mathscr{H}_{reson1}$ is kept.

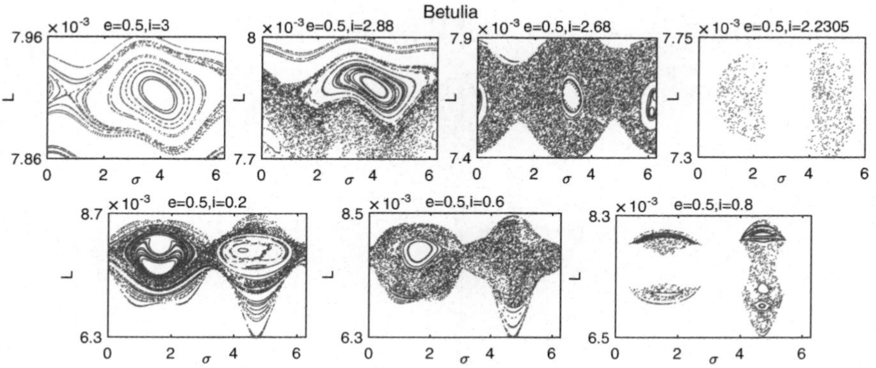

Fig. 5.17 The phase space of $\mathscr{H}_{0/2dof}$ at $e = 0.5$ for $i = 171.9°, 165°, 153.6°, 132°, 11.5°, 34.4°, 45.8°$

The Effect of e

In addition, it is also found that e shows the same effect on the dynamics of Betulia as for Vesta, both for the prograde and retrograde orbits. The mechanism is the same and is not explained in detail here. However, the phase space of $\mathscr{H}_{2dof}$ at $e = 0.5$ is included in Fig. 5.17. For the first plot at $i = 171.9°$, the main island is highly distorted although still without chaos. For $i = 165°$, the distortion is more serious and large chaos appears. The phase space with a small area of regular region in the center already shows the property of $\mathscr{H}_{reson2}$ at $i = 153.6°$. For $i = 11.5°, 34.4°, 45.8°$, the chaotic region is extended and new structures are generated.

5.5 The Maximal Lyapunov Characteristic Exponent of Chaotic Orbits

In addition to the above study about the extent of chaotic layers, the chaos can also be characterized quantitatively by calculating the value of the maximal Lyapunov Characteristic Exponent (mLCE), which is an indicator of the regular or chaotic properties of orbits [17]. Its basic idea is to measure the distance between two orbits that start close, until the infinite time $t \to \infty$. It characterizes the average growth rate of a small perturbation of the solution of a dynamical system and is defined as

$$\lambda = \lim_{t \to \infty} \frac{1}{t} \sum_{0}^{t} \ln \parallel \boldsymbol{v}(t) \parallel \tag{5.8}$$

in which $\boldsymbol{v}(t)$ is the deviation vector with respect to the given orbit at time t. It is also the solution of the corresponding variational equations of the dynamical system. If

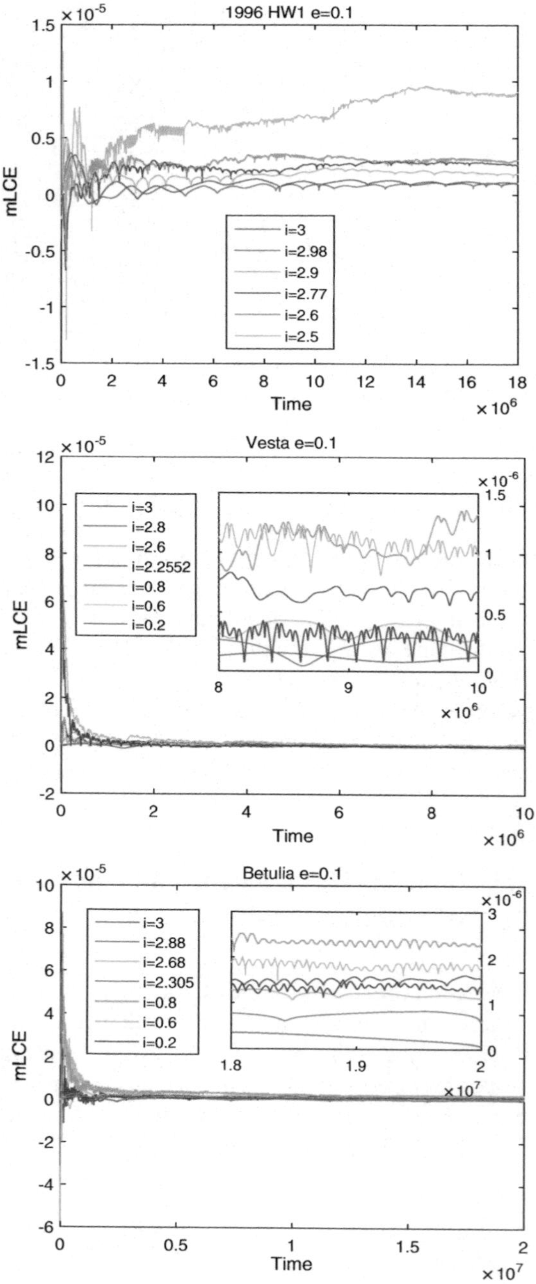

Fig. 5.18 The mLCE of regular and chaotic orbits from the Poincaré maps of $\mathscr{H}_{0/2dof}$ for 1996 HW1, Vesta and Betulia

$\lambda > 0$, the orbit is chaotic; if $\lambda = 0$, the orbit is regular. The numerical algorithm applied here is the standard method originally developed by [1]. Its detailed implementation can be found in [17]. It has to be mentioned that for regular orbits it might take a long time for λ to achieve zero. However, within a moderate time interval the tendency to zero is already visible. Since it is obvious that large e introduces stronger chaos and the chaos of the three different asteroids is expected to be compared, the mLCE of orbits selected from the chaotic and regular regions (if there is no chaos) on the maps from Figs. 5.5, 5.9, 5.10, 5.12 and 5.13 are given in Fig. 5.18. These maps primarily indicate the effect of i on dynamics at $e = 0.1$. The total integration time for them is different, but has been chosen such that a stable value of all the mLCE can be achieved. To make the results more visible, the mLCE at the end of the integrations are magnified and are shown respectively as insets in the plots of Vesta and Betulia.

For the three asteroids, they share the same property that the more inclined the orbit, the larger mLCE value it has, indicating the stronger chaotic property. In addition, the mLCE values of the retrograde orbits are generally smaller than those of the prograde ones. For $i = 171.9°$, there is no chaos on both the maps of Vesta and Betulia (shown in Figs. 5.5 and 5.9). This is demonstrated in the value for the mLCE illustrated as dark blue at the bottom of inset which will finally tends to zero, the tendency of which can already be identified. The difference among the three asteroids can also be noticed. The resonant orbits around 1996 HW1 have the largest mLCE (at magnitude 10^{-5}), the ones Betulia rank second (at magnitude 10^{-6}), while orbits around Vesta show the smallest mLCE (mostly at magnitude 10^{-7}). This can be explained by the different values of C_{20} and C_{22} induced from the irregular shape of the body. The more irregular the gravitational field is, namely 1996 HW1 in our study, the relatively larger C_{20} and C_{22} and the resultant larger perturbation from $\mathscr{H}_{reson2}$ it generates. Vesta is the most regular body and the influence of its second degree of freedom dynamics is limited. Betulia is in between.

In summary, the mLCE not only can identify the chaotic behaviour of orbits, but also gives us a hint on their extent of chaotic property.

5.6 Conclusions

In this study, a 2-DOF Hamiltonian of the 1:1 resonant dynamics of a gravitational field up to degree and order 4 was built. The 1-DOF Hamiltonian $\mathscr{H}_0$ was first studied by finding the EPs and examining their stability for non-circular and non-polar orbits of Vesta, 1996 HW1 and Betulia. This $\mathscr{H}_0$ was proven to capture the main characteristics of the 1:1 resonant dynamics, which was proved by the three study cases. For $\mathscr{H}_0$, i was found to play a significant role on the number of EPs. When i approaches π, there is only one stable EP left, due to the dominant strength of C_{31} over C_{22} on the structure of the phase space. The 2nd degree and order harmonics largely determine the stability of the EP, while the higher order terms either introduce new EPs and change the resonance width or break the symmetry of the dynamics.

By applying Poincaré maps, the 2-DOF Hamiltonian $\mathscr{H}_{2dof}$ was then investigated. Two resonances $\mathscr{H}_{reson1}$ and $\mathscr{H}_{reson2}$ were defined in this 2-DOF model and their locations and widths were determined numerically for different combinations of e and i.

With the overlap criteria, the extent of chaotic regions was qualitatively explained by the distance between the two resonances as well as their resonance strength. For small e and i close to 0 or π, the dynamics of $\mathscr{H}_{reson1}$ around the stable EP is hardly influenced for the situation when $\mathscr{H}_{reson1}$ and $\mathscr{H}_{reson2}$ are further apart. When i gets a bit further away from the equatorial plane, $\mathscr{H}_{reson2}$ becomes closer to and almost interacts with $\mathscr{H}_{reson1}$. Small-scale chaos (chaotic layers) were generated in the vicinity of the separatrix of $\mathscr{H}_{2dof}$, whose boundaries were well estimated by the modulated-pendulum approximation. When the two resonances have an obvious overlap for i getting close to the polar region, large chaos became apparent and new islands came forth in the phase space. However, for the near polar case, the libration region of $\mathscr{H}_{reson1}$ is hardly influenced and is stable against perturbation of $\mathscr{H}_{reson2}$. Though the structure of the phase space is largely determined by i, the large e definitely gives rise to strong perturbation of $\mathscr{H}_{reson2}$, which makes the main island distorted and the chaotic region extended. Therefore, the retrograde, near polar and near circular orbits show more stability against external perturbations.

In addition, the mLCEs of the chaotic and regular orbits were calculated, from which the above conclusion was proved quantitatively and the chaotic orbits around 1996 HW1 were revealed to have the strongest chaos, due to its highly irregular gravitational field.

The results and analyses in this chapter serve as an example of studying the relationship among resonance overlap, extent of chaos and strength of the perturbing terms. The 2-DOF resonant dynamics of other main motion resonances, e.g. 1:2, 2:3, 3:2, can also be investigated with the approach developed in this chapter.

References

1. G. Benettin, L. Galgani, Lyapunov characteristic exponents and stochasticity. *Intrinsic Stochasticity in Plasmas*, 17–23 June 1979
2. B.V. Chirikov, A universal instability of many-dimensional oscillator systems. Phys. Rep. **52**(5), 263–379 (1979)
3. A. Compère, A. Lemaître, N. Delsate, Detection by MEGNO of the gravitational resonances between a rotating ellipsoid and a point mass satellite. Celest. Mech. Dyn. Astron. **112**(1), 75–98 (2012)
4. N. Delsate, Analytical and numerical study of the ground-track resonances of Dawn orbiting Vesta. Planet. Space Sci. **59**(13), 1372–1383 (2011)
5. F. Delhaise, J. Henrard, The problem of critical inclination combined with a resonance in mean motion in artificial satellite theory. Celest. Mech. Dyn. Astron. **55**(3), 261–280 (1993)
6. J. Feng, R. Noomen, J. Yuan, Orbital motion in the vicinity of the non-collinear equilibrium points of a contact binary asteroid. Planet. Space Sci. **117**, 1–14 (2015)
7. J. Henrard, A semi-numerical perturbation method for separable Hamiltonian systems. Celest. Mech. Dyn. Astron. **49**(1), 43–67 (1990)

8. W. Hu, D.J. Scheeres, Numerical determination of stability regions for orbital motion in uniformly rotating second degree and order gravity fields. Planet. Space Sci. **52**(8), 685–692 (2004)
9. W.M. Kaula, *Theory of Satellite Geodesy: Applications of Satellites to Geodesy* (Courier Corporation, New York, 2000)
10. C. Magri, S.J. Ostro, D.J. Scheeres, M.C. Nolan, J.D. Giorgini, L.A.M. Benner, J.-L. Margot, Radar observations and a physical model of Asteroid 1580 Betulia. Icarus **186**(1), 152–177 (2007)
11. C. Magri, E.S. Howell, M.C. Nolan, P.A. Taylor, Y.R. Fernández, M. Mueller, R.J. Vervack, L.A.M. Benner, J.D. Giorgini, S.J. Ostro et al., Radar and photometric observations and shape modeling of contact binary near-Earth Asteroid (8567) 1996 HW1. Icarus **214**(1), 210–227 (2011)
12. A. Morbidelli, *Modern Celestial Mechanics: Aspects of Solar System Dynamics* (Taylor & Francis, London, 2002)
13. Ø. Olsen, Orbital resonance widths in a uniformly rotating second degree and order gravity field. Astron. Astrophys. **449**(2), 821–826 (2006)
14. T. Prieto-Llanos, M.A. Gomez-Tierno, Station keeping at libration points of natural elongated bodies. J. Guid. Control Dyn. **17**, 787–794 (1994)
15. D.J. Scheeres, Dynamics about uniformly rotating triaxial ellipsoids: applications to asteroids. Icarus **110**(2), 225–238 (1994)
16. D.J. Scheeres, S.J. Ostro, R.S. Hudson, R.A. Werner, Orbits close to asteroid 4769 Castalia. Icarus **121**(1), 67–87 (1996)
17. Ch. Skokos. The Lyapunov characteristic exponents and their computation. *Dynamics of Small Solar System Bodies and Exoplanets* (2010), pp. 63–135
18. S. Tzirti, H. Varvoglis, Motion of an Artificial Satellite around an Asymmetric, Rotating Celestial Body: Applications to the Solar System. Ph.D. Dissertation, Aristotle University of Thessaloniki, 2014
19. S. Valk, A. Lemaitre, F. Deleflie, Semi-analytical theory of mean orbital motion for geosynchronous space debris under gravitational influence. Adv. Space Res. **43**, 1070–1082 (2009)

Chapter 6
Orbital Dynamics in the Vicinity of Contact Binary Asteroid Systems

6.1 Summary

Many studies about orbital dynamics around single asteroids and binary asteroid systems have been performed. With this background, the research in this chapter focuses on orbital dynamics in the vicinity of contact binary asteroids, which is estimated to constitute 10–20% of all small solar system bodies (including comets). This kind of bodies is also characterized by their highly bifurcated shape. In August 2014, ESA's Rosetta arrived at its target comet 67P/Churyumov-Gerasimenko after a ten-years journey in space. The comet, impressing many by virtue of its shape with two lobes in contact, belongs to contact binary bodies. In addition, in October 2015, the images sent back by NASA's New Horizons revealed that Pluto's tiny moon Kerberos also consists of two lobes. These discoveries provide good support for the selection of this topic. In addition, this characterization also can give us hints on the formation and evolution of such small solar system bodies. Specifically, the objective of this research is to perform a systematic study on the orbital motion around contact binary asteroid systems.

For the purpose of this study, first, the gravitational field needs to be modeled. There are three main methods: (1) the spherical harmonics expansion; (2) the polyhedron model that approximates the body with large numbers of polyhedra, given that the detailed shape model of the body is available; (3) geometric shape models whose closed-form potentials can be usually obtained. To fulfill the purpose for a systematic study about the dynamical environment around contact binary bodies, the spherical harmonics and configuration of a combination of a sphere and an ellipsoid are applied, respectively.

Firstly, the ellipsoid-spherical configuration is applied. The gravitational field can be obtained in the closed-form potential from the two components and the equations of motion can be built in the rotating frame. Actually, this model is analogous to that of the CR3BP, with the same common definition of the mass ratio, which is the ratio of mass of the smaller component to the total mass of the two main bodies. However, an important difference is that the ratio of gravitational acceleration to

J. Yuan et al., *Low Energy Flight: Orbital Dynamics and Mission Trajectory Design*, https://doi.org/10.1007/978-981-13-6130-2_6

centrifugal acceleration equals one for the two main bodies of the CR3BP, but it can be any number for our model. For the situation where it is larger than one, there is compression between the two lobes, and stretching occurs for ratios smaller than one. The value of one indicates that there is no internal force between the two. Therefore, our model can be viewed as a generalization of the CR3BP. By varying the values of these two ratios in our model, contact binary asteroids with this kind of shape and with a wide range of physical parameters can be covered. Due to this similarity, the methods that apply to the CR3BP for characterizing orbital motion are obtained and examined for the first time in application to contact binary asteroid system 1996 HW1, which is known to be the most bifurcated asteroid.

Accordingly, equilibrium points (EPs) and their stability are first obtained and examined for HW1. By varying the values of the two ratios in the model, their influence on the location and stability of the EPs is examined. Second, Lyapunov, Halo and vertical (the motion is mostly along the z-direction) families of periodic orbits (POs) are obtained for HW1. The fast rotation of the asteroid has a stabilizing effect on equatorial orbital motion. Orbits that are in resonance with the rotation of the asteroid are also obtained, and they can provide good coverage of the polar region of the body for spacecraft observations.

In the CR3BP, there is a well-known Richardson third-order analytical solution of Halo orbits obtained with the Lindstedt-Poincaré (LP) method. In this research, the same method is applied to obtain the third-order analytical solution of orbital motion in the vicinity of the non-collinear EPs with non-spherical gravitational field. This solution is tested against numerical simulations and it proves to have a very good accuracy for moderate-amplitude orbital motion. With the increase of orbital amplitude and rotation rate of the asteroid, the solution becomes less accurate, due to the application of linear expansion of the LP method.

Finally, the methods developed in this chapter can also be applied to the study of orbital dynamics in binary asteroid systems, planet-moon systems and binary star systems.

6.2 Numerical Analysis of Orbital Motion Around Contact Binary Asteroid System

The general orbital motion around a contact binary asteroid system is investigated in this study. System 1996 HW1 is explored in detail, as it is the mostly bifurcated asteroid known to date. The location of its equilibrium points (EPs) is obtained and their linear stability is studied. Families of Lyapunov, Halo and vertical periodic orbits (POs) in the vicinity of these EPs as well as their stability are found and examined, respectively. The influence of the relative size of each lobe and the shape of the ellipsoidal lobe and the rotation rate of the asteroid on the location and stability of the EPs are studied. Additionally, two families of equatorial orbits are obtained at a wide range of distances: from far away to nearby. Their stability is examined against the distance to the asteroid and the rotation rate of the asteroid, to uncover the influence of irregularities in the gravitational field and the rotation of the asteroid on the orbital motion. Finally, resonant orbits in N commensurability with the rotation of the asteroid are found and their stability is discussed. The fast rotation of the asteroid has a stabilizing effect on the equatorial orbital motion.

6.2.1 Introduction

Up to now, several space missions destined for small solar system bodies, e.g. asteroids and comets, have been launched. Close proximity operations are challenging for these missions due to perturbations caused by the irregular gravitational fields of these bodies. Comet 67P/Churyumov-Gerasimenko, the target of ESA's Rosetta mission, was found to be a contact body of two lobes with different origination recently [50]. NASA's New Horizon mission discovered that one of Pluto's tiny moons Kerberos is also a double-lobed body. From radar and optical observations, many near-Earth asteroids (NEAs), main-belt and Trojan asteroids, and even comets are found to be contact binaries; they are estimated to constitute 10–20% of all small solar system bodies [16]. This study focuses on investigating the general properties of orbital motion in the strongly perturbed environment induced by these highly bifurcated bodies.

Traditionally, the shape of an asteroid was approximated by a triaxial or oblate ellipsoid. With this model and the closed-form ellipsoidal potential, Chauvineau et al. [6] investigated planar orbits by numerical integration and identified chaotic and regular orbits by varying the mass distribution and rotation rate of the body. Scheeres [44] performed systematic studies about the EPs and POs in its vicinity, from which an asteroid was classified as type I if the non-collinear EPs are stable and type II if they are unstable. Werner [54] developed the polyhedron method, to approximate the shape and gravitational field of asteroids by means of thousands of polyhedra. This is the most accurate approach, especially for studying motion extremely close to and on the surface of an asteroid. It has been widely applied for identifying the dynamical environment, e.g. EPs, POs and particle motions, around asteroids with detailed shape models [47–49]. A closely related model is the so-called 'mascons' model that represents the asteroid with a collection of point masses, which was first used to estimate the lunar gravitational potential [36]. However, it is less accurate on the surface of the body, compared to the polyhedron model [55]. The spherical harmonics model was also widely applied for general analytical and averaging studies of orbital motion around asteroids. From this model, the C_{20}, C_{30} and C_{40} terms were found to introduce secular rates of the argument of periapsis, the ascending node, mean anomaly and eccentricity for orbits close to the asteroid. The C_{22} term was identified to change orbital energy and angular momentum [46]. Even with higher degree and order spherical harmonics, frozen orbits were obtained [5]. In addition, the geometrical shapes of a cube, a straight segment, and two orthogonal

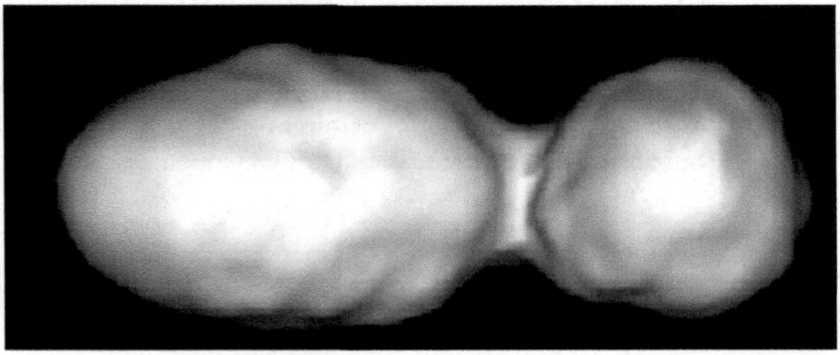

Fig. 6.1 The shape model of 1996 HW1 [31]

segments were also applied [1, 27, 40]. Families of POs were generated in their vicinity and the stability of the orbits was investigated. The region for stable orbital motion around an asteroid represented by a straight segment was identified in Lara and Scheeres [25].

For our exploration of the dynamical environment around contact binary bodies, a shape model consisting of two lobes (an ellipsoidal component and a spherical component) that are in physical contact, is applied. Based on it, the effects of the system configuration (varying the relative size of each lobe and the shape of the ellipsoidal lobe) and the rotation rate of the asteroid on the orbital motion can be studied in a systematic way. For this kind of shape, Scheeres [45] discussed formation mechanisms and studied the relationship between the relative configuration and the rotational angular momentum. The motion of an orbiting spacecraft or a particle in its vicinity will be investigated in detail in this study. Here, the ellipsoid and the sphere are combined in one body, which breaks the symmetry along one axis of the system. This is different from the previous models that approximate the bifurcated body by two connected spheres [11] and two mass dipoles [39], which have complete symmetry in three axes.

It is known from Magri et al. [31] that contact binary system 1996 HW1 is the mostly bifurcated body (asteroid) ever found (probably surpassed by comet 67P observed by the Rosetta mission), as shown in Fig. 6.1. Therefore, it serves as the basic model of a contact binary asteroid and its physical parameters are applied in our simulations. Although there are several methods to represent the gravitational field of this highly bifurcated shape, e.g. the spherical harmonics expansion and the polyhedron approximation as mentioned, in this study the potential from the combination of an ellipsoidal potential and a spherical potential is directly applied, with the constant-density assumption. Since the orbital motion in the vicinity of a highly perturbed gravitational field is the focus, other perturbations, e.g. solar radiation pressure and third-body gravitation (e.g. solar gravitation), are not considered here. For this kind of shape model, there are two free system parameters: the mass ratio that reflects the mass distinction of the two components and the gravitational-centripetal acceleration ratio δ that indicates the rotation situation (fast or slow) of the asteroid. Therefore, this study is arranged as follows. First, based on the physical parameters of the system 1996 HW1, the EPs and their linear stability are identified. Second, the influence of μ and δ on the location and stability of the EPs is investigated. Third, with approximated analytical initial conditions and the differential correction (DC) method and the continuation process, families of Lyapunov, Halo and vertical POs are generated, and their stability is studied. In addition, equatorial POs around the entire system and their linear stability are investigated, and the effect of the parameter δ on their stability is examined. Finally, resonant orbits that are in N commensurability with the rotation of the asteroid, are obtained for different δ, and their stability is discussed.

6.2.2 Dynamical Model

The geometry of the ellipsoid-sphere configuration is illustrated in Fig. 6.2. The parameters that characterize this configuration are: the three semi-axes of the ellipsoid α, β, γ, the radius of the sphere R and the uniform rotation rate ω. The system is assumed to be homogeneous, with a constant density ρ. The vector between the centers of mass of the two components is defined to be $\boldsymbol{d}$ (from ellipsoid to sphere), where $|\boldsymbol{d}| = \alpha + R$, and the mass ratio μ is equal to $m_s/(m_s + m_e) = R^3/(R^3 + \alpha\beta\gamma)$ (m_s and m_e being the mass of the sphere and the ellipsoid, respectively). In the body-fixed frame ($XYZ-$frame in Fig. 6.2), the gravitational potential is invariant, and the equations of motion for an object located at $\boldsymbol{r} = (x, y, z)$ in the vicinity of the asteroid can be written as

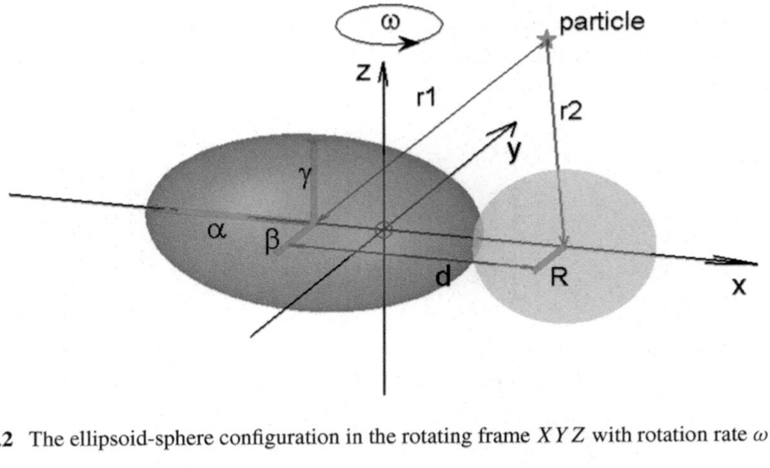

Fig. 6.2 The ellipsoid-sphere configuration in the rotating frame XYZ with rotation rate ω

$$\ddot{\boldsymbol{r}} + 2\boldsymbol{\omega} \times \dot{\boldsymbol{r}} + \boldsymbol{\omega} \times (\boldsymbol{\omega} \times \boldsymbol{r}) = \frac{\partial U_{se}}{\partial \boldsymbol{r}} \tag{6.1}$$

where the potential of the asteroid U_{se} is expressed as

$$U_{se} = U_s + U_e = G(m_s + m_e)\left(\frac{\mu}{|\boldsymbol{r} - (1-\mu)\boldsymbol{d}|} + (1-\mu)U_E(\boldsymbol{r} + \mu\boldsymbol{d})\right) \tag{6.2}$$

in which U_s and U_e are the potential of the spherical lobe and ellipsoidal lobe, respectively, and U_E is the ellipsoid potential of unit mass and is expressed as [29]

$$\begin{aligned} U_E(\boldsymbol{s}) &= \frac{3}{4}\int_\lambda^\infty \phi(\boldsymbol{s}, v)\frac{dv}{\Delta(v)} \\ \phi(\boldsymbol{s}, v) &= 1 - \frac{x^2}{\alpha^2 + v} - \frac{y^2}{\beta^2 + v} - \frac{z^2}{\gamma^2 + v} \\ \Delta(v) &= \sqrt{(\alpha^2 + v)(\beta^2 + v)(\gamma^2 + v)} \end{aligned} \tag{6.3}$$

where $\phi(\boldsymbol{s}, v) = 0$, $\boldsymbol{s} = (x, y, z)$. Taking the length and time units as the distance d and ω^{-1}, respectively, after normalization, Eq. (6.1) becomes

$$\begin{cases} \ddot{x} - 2\dot{y} = x - \delta\left[\dfrac{\mu(x - 1 + \mu)}{|\boldsymbol{r}_2|^3} - \dfrac{(1-\mu)\partial U_E(\boldsymbol{r}_1)}{\partial x}\right] \\ \ddot{y} + 2\dot{x} = y - \delta\left[\dfrac{\mu y}{|\boldsymbol{r}_2|^3} - \dfrac{(1-\mu)\,\partial U_E(\boldsymbol{r}_1)}{\partial y}\right] \\ \ddot{z} = -\delta\left[\dfrac{\mu z}{|\boldsymbol{r}_2|^3} - \dfrac{(1-\mu)\,\partial U_E(\boldsymbol{r}_1)}{\partial z}\right] \end{cases} \tag{6.4}$$

where $\boldsymbol{r}_1$ and $\boldsymbol{r}_2$ are the vectors from the particle (x, y, z) to ellipsoid center $(-\mu, 0, 0)$ and the sphere center $(1-\mu, 0, 0)$, respectively, and $\delta = G(m_s + m_e)/\omega^2 d^3$ is a dimensionless scaling parameter that represents the ratio of the gravitational acceleration to centrifugal acceleration. The derivatives of the ellipsoid potential with respect to x, y, z are

$$\begin{cases} U_{Ex} = -\frac{3}{2}x\int_\lambda^\infty \frac{dv}{(\alpha^2+v)\Delta(v)} \\ U_{Ey} = -\frac{3}{2}y\int_\lambda^\infty \frac{dv}{(\beta^2+v)\Delta(v)} \\ U_{Ez} = -\frac{3}{2}z\int_\lambda^\infty \frac{dv}{(\gamma^2+v)\Delta(v)} \end{cases} \tag{6.5}$$

in which the parameters $x, y, z, \alpha, \beta, \gamma$ are all normalized, and the derivatives can be computed using the Carlson integral of the second kind [38]. The second derivatives of U_E with respective to x, y, z can be found in [46]. Therefore, for orbital motion in the vicinity of the asteroid, the potential U_E can be rewritten as

$$U_E = \frac{3}{2} \cdot R_0 + \frac{1}{2}(x \cdot U_{Ex} + y \cdot U_{Ey} + z \cdot U_{Ez}) \tag{6.6}$$

where

$$R_0 = \frac{1}{2} \int_0^{\infty} \frac{dv}{\Delta (v + \lambda)}$$

is the Carlson integral of the first kind [38]. The integral for this Hamiltonian system is

$$C = \frac{1}{2}\left(\dot{x}^2 + \dot{y}^2 + \dot{z}^2\right) - \frac{1}{2}\left(x^2 + y^2\right) - \delta U_{se} = T - \mathrm{V} \tag{6.7}$$

in which $T = (\dot{x} + \dot{y} + \dot{z})/2$ is the specific kinetic energy, $V = (x^2 + y^2)/2 + \delta U_{se}$ is the effective potential, and C is the Jacobi integral or Jacobi constant.

The free parameters of Eq. (6.4) are μ and δ. The former one reflects the mass distribution within the system and the relative size (or mass) of the two components. The latter one indicates the rotation situation of the asteroid. Parameter $\delta = 1$ represents the case that the two lobes are just touching one another without any internal forces, while the components are in compression for $\delta > 1$ and stretch for $\delta < 1$. It is pointed out here that the CRTBP can be viewed as a particular situation in which both of the two lobes are spheres with $\delta = 1$. The influence of μ and δ on the dynamical environment in the vicinity of the asteroid will be investigated in the following sections.

6.2.3 *Contact Binary System 1996 HW1*

Firstly, numerical studies are performed for 1996 HW1, whose detailed shape model was obtained by Magri [31] (Fig. 6.1). It is found to be the most bifurcated bodies among the currently known elongated asteroids, with a pronounced 'neck' separating two lobes in a roughly 1:2 mass ratio. Since the two components can be represented by an ellipsoid and a sphere, Table 6.1 gives their physical dimensions, the bulk density and the rotational period of the entire system. Given these rotational period and the bulk density, the system is not in a minimum energy state as described in [45] as δ is 2.1682, implying the existence of internal tension between the two components.

6.2.3.1 Zero-Velocity Curves and EPs

Similar to the CR3BP, for our model $C = -V$ also defines the zero-velocity surfaces with $T = 0$, which divides the space into an accessible region ($T > 0$) and a forbidden region ($T < 0$). Unlike the RTBP where the zero-velocity curves are largely determined by the mass ratio μ, the curves of our model are determined by μ together with δ. Given a value of the Jacobi constant C, the motion is bounded

Table 6.1 The main physical parameters of 1996 HW1 [31]

Overall dimensions (km)	$X: 3.78 \pm 0.05$; $Y: 1.64 \pm 0.1$; $Z: 1.49 \pm 0.15$
Sidereal period (h)	8.76243 ± 0.00004
Average sphere radius (km)	1.5
Triaxial ellipsoid principal axes size (km)	$2.28 \times 1.64 \times 1.49$
Bulk density ($g \cdot \mathrm{cm}^{-3}$)	2.0

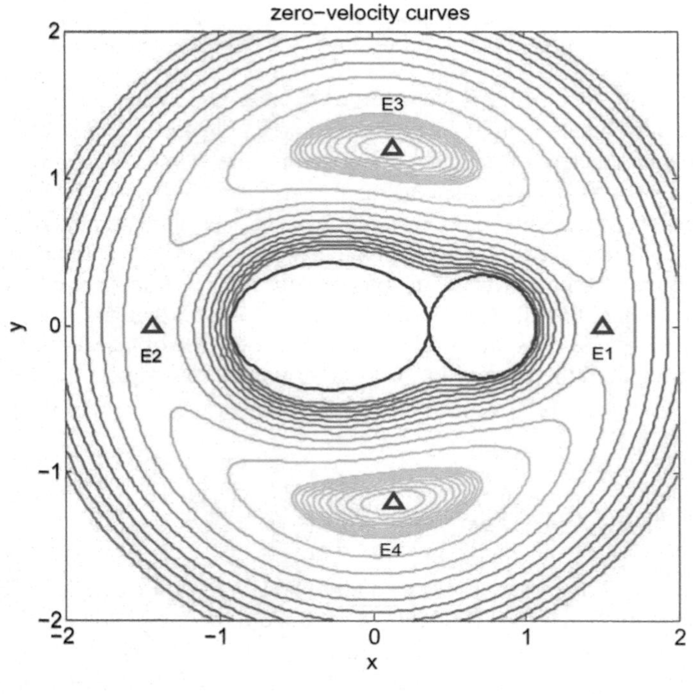

Fig. 6.3 The zero-velocity curves and positions of four EPs of system 1996 HW1 in the xy-plane

to the accessible region. In the $x - y$ plane, the zero-velocity curves for different C levels of system 1996 HW1 are shown in Fig. 6.3.

EPs are fixed points in the rotating (body-fixed) frame, where velocity and accelerations are zero. They can be interpreted as synchronous orbits around the asteroid in the inertial frame. In Fig. 6.3, four EPs can be identified, among which E1, E2 are the collinear ones and E3, E4 are the non-collinear ones. According to [44], E1 and E2 are referred to as saddle EPs, while E3 and E4 are the so-called center EPs. As this configuration is symmetric in the z-plane, the EPs are all situated in the $x - y$ plane. Their exact positions can be computed by setting the right-hand sides of Eq. (6.4) to zero and are denoted as $(x_e, y_e, z_e)^T$ According to differential theory

[32], the linear stability of the EPs can be determined from the linearized dynamics of Eq. (6.4). Given a small perturbation from an EP $\boldsymbol{X} = (\xi, \eta, \zeta, \dot{\xi}, \dot{\eta}, \dot{\zeta})^T$, in which $(\xi, \eta, \zeta)^T = (x, y, z)^T - (x_e, y_e, z_e)^T$, the linearized equation of motions for a perturbation (variational equations) in the neighborhood of an EP is written as

$$\dot{\boldsymbol{X}} = \boldsymbol{A} \cdot \boldsymbol{X} \tag{6.8}$$

where

$$\boldsymbol{A} = \begin{bmatrix} \mathbf{0}_{3\times 3} & \mathbf{I}_{3\times 3} \\ \begin{matrix} V_{xx} & V_{xy} & V_{xz} \\ V_{xy} & V_{yy} & V_{yz} \\ V_{xz} & V_{yz} & V_{zz} \end{matrix} & \begin{matrix} 0 & 2 & 0 \\ -2 & 0 & 0 \\ 0 & 0 & 0 \end{matrix} \end{bmatrix}$$

is the Jacobi matrix calculated at the corresponding EP and V is the effective potential defined in Eq. (6.7). The characteristic equation of $\boldsymbol{A}$ can be written as

$$\lambda^6 + a\lambda^4 + b\lambda^2 + c = 0$$

in which λ is the eigenvalue and coefficients a, b, c are determined by the second derivatives of V that are closely related to the rotation rate of the body. If all the eigenvalues have non-positive real parts, the EP is Lyapunov stable (or linearly stable), otherwise it is unstable. The positions of the four EPs of 1996 HW1 and their corresponding eigenvalues are given in Tables 6.2 and 6.3.

Since both E1 and E2 have positive real eigenvalues, they are hyperbolically unstable. E3 and E4 have eigenvalues with positive real parts, and they are complex unstable. The positions and stability of the EPs obtained here are highly consistent with those obtained by Magri [31], in which study the polyhedron model of the

Table 6.2 Location and linear stability of the EPs in the rotating frame

EP	x	y	Stability
E1	1.50397208867676	0	U
E2	−1.43907984894912	0	U
E3	0.142251271693655	1.20262697830487	U
E4	0.142251271693655	−1.20262697830487	U

Table 6.3 The eigenvalues of the EPs

EP	$\lambda_{1,2}$	$\lambda_{3,4}$	$\lambda_{5,6}$
E1	± 1.15329441819126	$\pm 1.327198177844053i$	$\pm 1.252450802130986i$
E2	± 0.90255553930741	$\pm 1.21107228063561i$	$\pm 1.16099544900511i$
E3/E4	-0.480938988379065 $\pm 0.852439624239106i$	$0.480938988379066 \pm$ $0.852439624239106i$	$\pm 1.004638240930704i$

body was used. This proves the validity of our model that approximates the body with the combination of an ellipsoid and a sphere. In addition, system 1996 HW1 can be classified as a type II asteroid, according to Scheeres [44]. The eigenvalues and eigenvectors determine the orbital motion in the vicinity of the EPs. For E1 and E2, the pair of real eigenvalues and their corresponding eigenvectors define the 1-dimensional stable and unstable manifolds. For E3 and E4, the two pairs of complex eigenvalues and eigenvectors define the spiral stable and unstable manifolds. The pure imaginary eigenvalues and the corresponding eigenvectors of all four EPs generate two 2-dimensional center manifolds, on which POs can be found. These POs are obtained in the following section.

6.2.3.2 Location and Stability of EPs at Different Values of μ and δ

Having obtained the EPs of system 1996 HW1 and their stability, a systematic study on the effects of μ and δ on the location and stability of the EPs is carried out for the same shape model of two connected lobes. As is closely related to the configuration of the system, Table 6.4 gives its value for different dimensions of the ellipsoid and the sphere, respectively, assuming a homogenous density. The radius of the sphere component is varied from 1 to 0.25 km, while the ellipsoid component changes from a sphere (A) to an ellipsoid (B) and then to a more elongated one (C). Therefore, the combinations of them include the sphere-sphere and ellipsoid-sphere configurations. The sphere-sphere configuration includes the case of two equal-sized spheres but also two configurations with a big and a small sphere. These configurations cover almost all possible configurations of a contact binary asteroid system.

In addition to μ, the parameter δ ranges from the critical value to the value of 10 for each configuration. Here the critical value is the value for which at least one of the EPs is located on the surface of the asteroid [53]. The positions of the EPs are given in Fig. 6.4. For all the configurations, the EPs move further away from the asteroid as μ increases, i.e. the rotation slows down and the centripetal contribution becomes smaller. Since the sets of three lines for E1 and E2 tend to overlap, the location of them are slightly influenced by μ, compared to δ. However, the dependence of the location of E3 on μ is strong. On the other hand, for E3 and a given value of μ, with an increment of δ, the y coordinate increases while the x coordinate does not change much. In addition, δ does not have much influence on the stability of

Table 6.4 The mass ratio μ of contact binary asteroid systems with different configurations

Sphere Radius/km	Ellipsoid Semi-axis/km		
	A:1 × 1 × 1	B:1 × 0.75 × 0.5	C:1 × 0.5 × 0.25
1	0.5	0.7273	0.8889
0.5	0.1111	0.25	0.5
0.25	0.0154	0.04	0.1111

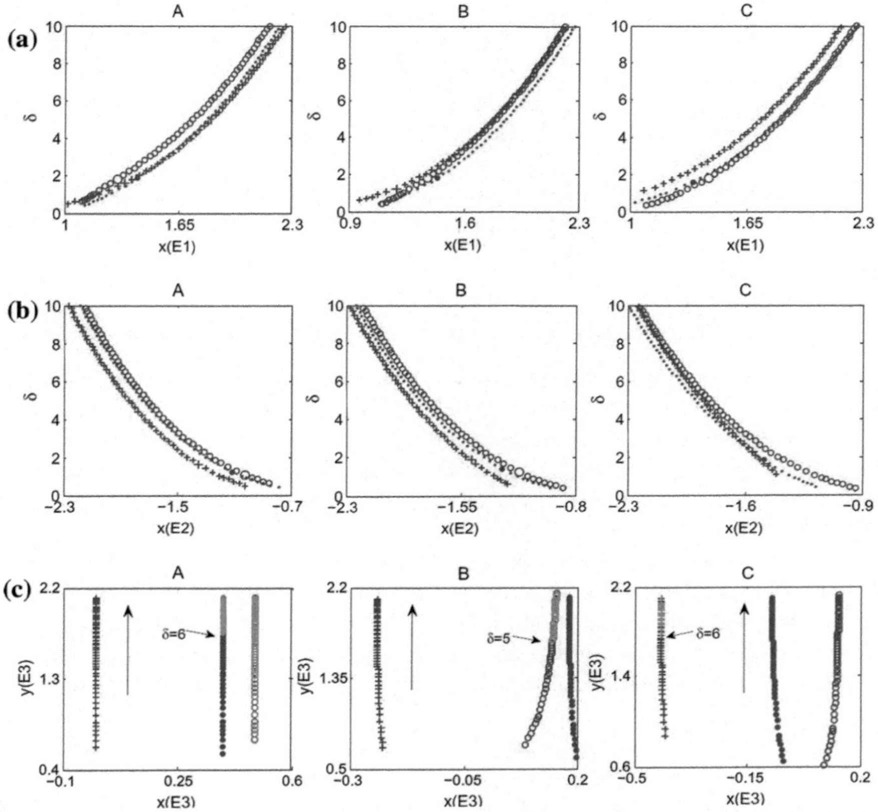

Fig. 6.4 The coordinates of E1 (upper), E2 (middle) and E3 (bottom) for the A, B and C configurations in Table 6.4, where the circles, dots and plus markers represent sphere components with radius of 0.25, 0.5 and 1 km, respectively. The blue and red dots indicate the unstable and stable EPs, respectively

E1 and E2, as both of them are always hyperbolically unstable. This is different for E3, which transits from linearly stable to unstable with the decrease of δ (fast rotation), as indicated in the bottom plots of Fig. 6.4. The red dots represent that E3 is linearly stable, while the blue ones indicate complex instability. The transitions always occur approximately at $\delta = 5$ or $\delta = 6$, which is worth further study on this specific value. Fast rotation (small δ) of the asteroid makes the E3 points unstable, which is consistent with the conclusion in [46].

6.2.3.3 POs in the Vicinity of the EPs

For the linearized system Eq. (6.8), provided that the initial conditions are restricted so that only the non-divergent mode is allowed, the 3-dimensional solution around the collinear EPs can be written as

$$\begin{cases} \xi = A_1 \cos \lambda t + A_2 \sin \lambda t \;\; (a) \\ \eta = -\alpha A_1 \sin \lambda t + \alpha A_2 \cos \lambda t \, (b) \\ \zeta = B_1 \sin \lambda_v t + B_2 \cos \lambda_v t \, (c) \end{cases} \tag{6.9}$$

in which λ and λ_v are the frequency of the motion in the $x - y$ plane and z-plane, respectively. After Moulton [35], the POs are classified into three categories

(1) $\xi = \eta = 0$ and ζ is in the form of (c).
(2) $\zeta = 0$ and ξ, η are in the form of (a), (b) respectively.
(3) ξ, η, ζ are of the form of (a), (b), (c) respectively, and are commensurable.

Each of these options will be discussed later. In addition, from the linearized system of the CR3BP, Szebehely [52] studied the possibility of short-period and long-period periodic motions around the EPs and their stability. Similarly, Lara and Elipe [23] investigated the linearized motion in the vicinity of the geostationary points and obtained families of planar periodic orbits. For our study, when E3 is stable for some rotation rates of the asteroid as studied in Sect. 6.2.3.2, there exists both short- and long-period motions in the vicinity of it.

According to the Floquet Theorem [32], the linear stability of POs can be determined from their State Transition Matrix (STM) $\boldsymbol{\Phi}(t)$, which reflects the change of state variables at epoch t due to a small deviation of the state at initial time. It is defined as a nonsingular matrix which satisfies the matrix differential equation

$$\dot{\boldsymbol{\Phi}}(t) = \boldsymbol{A} \cdot \boldsymbol{\Phi}(t)$$

where $\boldsymbol{A}$ is the matrix defined in the linearized system Eq. (6.8) and $\boldsymbol{\Phi}(0) = \boldsymbol{\Phi}(6 \times 6)$. For a PO with period T_p, the STM after completion of one full period is named the monodromy matrix $\boldsymbol{M}$, i.e. $\boldsymbol{M} = \boldsymbol{\Phi}(T_p)$. The eigenvalues of $\boldsymbol{M}$ are called Floquet multipliers and $\boldsymbol{M}$ was proven to be symplectic for the autonomous Hamiltonian system. Therefore, $\boldsymbol{M}$ has eigenvalues in the form of $\lambda_1, \frac{1}{\lambda_1}, \lambda_2, \frac{1}{\lambda_2}1, 1$. Following Gómez [12], the stability index here is defined as $s_i = |\lambda_i + \frac{1}{\lambda_i}|, i = 1, 2$. The PO is stable if $s_i < 2$, while unstable if $s_i > 2$. Bifurcations might occur and new families of POs are expected to be generated at $s_i = 2$. In the following study, only the pairs of non-unit eigenvalues are considered.

6.2.3.4 Lyapunov Orbits

The second category of Moulton's classification is considered first, which corresponds to planar motion. Therefore, the general solution of the linearized system can be written as

$$\begin{cases} \xi = A \sin(\lambda t + \phi) \\ \eta = \alpha A \cos(\lambda t + \phi) \end{cases} \tag{6.10}$$

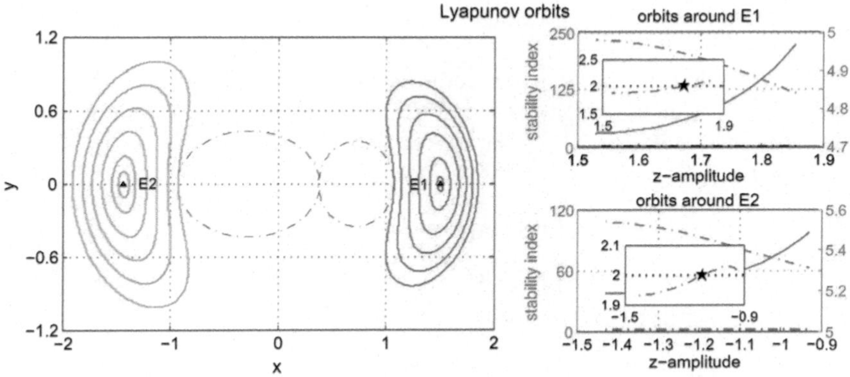

Fig. 6.5 Lyapunov orbits around E1 and E2; Right: their stability index s_1 and s_2 (blue dash-dotted lines) and corresponding periods (orange lines)

where λ is the mode of the first pair of pure imaginary eigenvalues of E1 and E2 (listed in Table 6.3), $\alpha = ((\lambda + V_{xx}/\lambda))/2$, and A is the amplitude. As our model is symmetric with respect to the x-axis in the $x-y$ plane, the initial condition of the PO is chosen as the point that intersects the x-axis perpendicularly:

$$x_0 = x_{E1} + A,\ y_0 = 0,\ \dot{x}_0 = 0,\ \dot{y}_0 = -\alpha\lambda A,\ T = 2\pi/\lambda$$

Based on the differential correction (DC) process that has been widely applied for finding POs around asteroids [49] and planetary moons [24, 42], these approximate initial conditions are adjusted and the exact solutions of the full non-linearized model are obtained. For obtaining large-amplitude orbits, the numerical continuation method [37, 42] is applied if the DC does not work well with the given initial conditions.

With the increase of orbital amplitude A, the Lyapunov orbits expand from the vicinity of the EP to the surface of the asteroid. Orbits around E1 and E2 that do not intersect with the asteroid are illustrated in Fig. 6.5, as well as the stability index and periods. It can be seen that all orbits are highly unstable with large s values. There are two blue dash-dotted lines in each stability plot. The upper one is s_1 that indicates the in-plane stability, while the lower one is s_2 that represents the vertical (or z-direction) stability, and $s_3 = 2$ is the pair of unit eigenvalues and is not shown in the plot. The closer the orbit to E1 and E2, the larger the instability is. For orbits around E1 and E2, their s_2 passes the critical line $s = 2$ at $x \approx 1.77$ and $x \approx -1.1$ (marked as pentagrams in the plot), respectively, which means that bifuracations occur and new families of POs are generated at these locations. This phenomenon also exists in the CR3BP, for which the new family orbits are actually the Halo orbits. Furthermore, for all these orbits, their periods become longer with an incresase of the orbital amplitudes.

6.2.3.5 Halo Orbits

The general 3-dimensional orbits around EPs are known as Lissajous orbits in the CR3BP. They are quasi-periodic. Halo orbits are a special classification where the frequency of the orbit in the $x - y$ plane is equal to that in the z-direction, i.e. $\lambda = \lambda_v$. Therefore, Eq. (6.9) can be written as [41]

$$\begin{cases} \xi = A_x \cos(\lambda t + \phi) \\ \eta = \alpha A_x \sin(\lambda t + \phi) \\ \varsigma = A_z \sin(\lambda t + \psi) \end{cases} \\ \psi = \phi + n\pi/2, n = 1, 3 \tag{6.11}$$

in which A_x and A_z are the amplitudes in the x-direction and z-direction, respectively. [41] derived the third-order analytical solution of the above system and gave solutions explicitly. This approximation has been primarily used in the CR3BP. However, in our current numerical study, we use the initial conditions that are directly obtained from Eq. (6.11) and are written as

$$x_0 = x_{E1} + A_x, y_0 = 0, z_0 = \pm A_z, \dot{x}_0 = 0, \dot{y}_0 = -\alpha\lambda A, \dot{z}_0 = 0, T = 2\pi/\lambda$$

The conditions A_z and $-A_z$ generate two families of Halo orbits, respectively, which are known as the northern Halo orbits and southern Halo orbits in the CRTBP [19]. By applying the same DC method, orbits around both E1 and E2 are obtained and the ones that do not intersect with the asteroid are depicted in Fig. 6.6. For a good visualization, their projections on the $x - y$, $x - z$ and $y - z$ planes are given in Fig. 6.7. It can be seen that the z-amplitude A_z controls the size and the shape of the orbits. The purple and green orbits are the so-called northern and southern

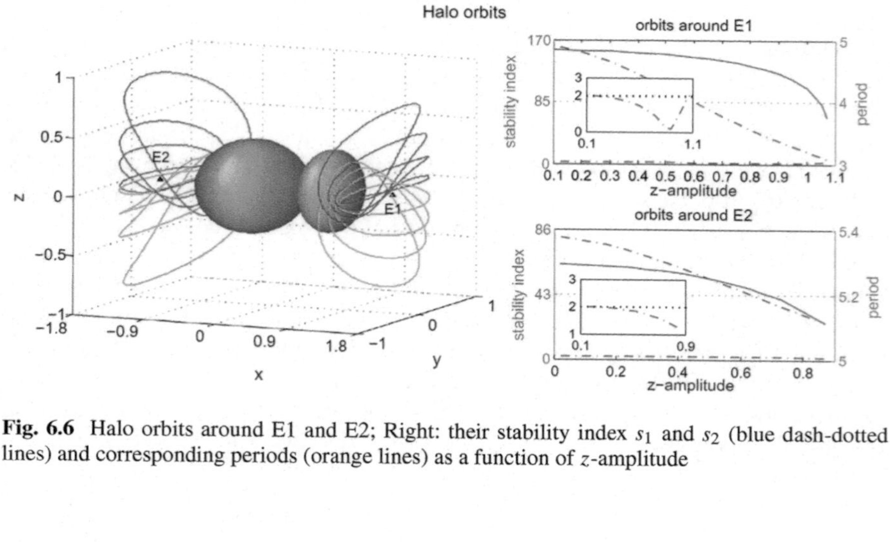

Fig. 6.6 Halo orbits around E1 and E2; Right: their stability index s_1 and s_2 (blue dash-dotted lines) and corresponding periods (orange lines) as a function of z-amplitude

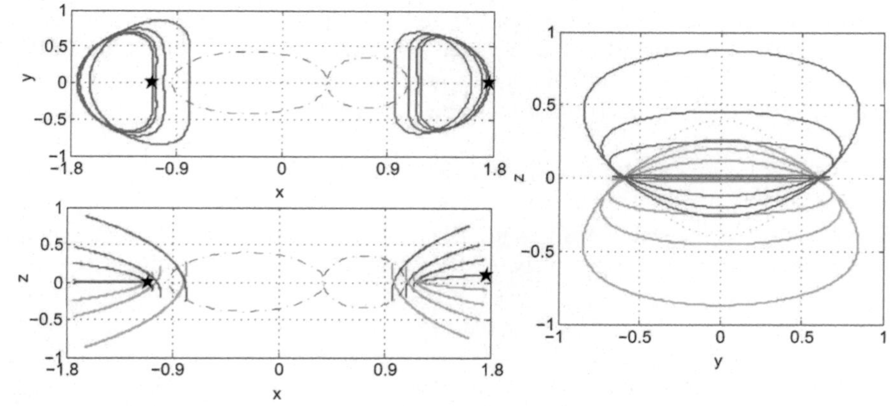

Fig. 6.7 The projections of the Halo orbits around E1 and E2 on the $xy-$, $xz-$ and $yz-$ planes

Halo orbits, respectively. When A_z increases, the orbits become more inclined and move further away from E1 and E2 but close to the surface of the asteroid. From the stability index, it can be found that the closer the orbit to E1 and E2 the larger its instability. These orbits all have stability in the z-direction, as shown by the fact that s_2 has values no larger than 2. However, there is a tendency that the orbit will become z-direction unstable if its A_z becomes larger. The orbital period reduces with the increase of A_z.

As shown in Fig. 6.7, the projections of Halo orbits on the $x-y$ plane have a similar shape with that of the Lyapunov orbits. For extremely small A_z, the orbits approach to the $x-y$ plane infinitely which is witnessed by the projection in the $x-z$ plane. They originate from the planar motion as their vertical stability starts from value 2, which is the bifurcation point. The x coordinates of the near planar Halo orbits (marked in pentagrams in Fig. 6.7) are approximately the same with those of the bifurcation points for the Lyapunov orbits in Fig. 6.5. Therefore, similar to the CRTBP, Halo orbits are actually bifurcations of the Lyapunov family orbits for our model.

6.2.3.6 Vertical Orbits

In this research, a vertical orbit is an orbit with its motion mostly in the z-direction. For this kind of orbits, the initial conditions from Eq. (6.9) are

$$\xi_0 = \eta_0 = \zeta_0 = 0, \dot{\xi}_0 = \dot{\eta}_0 = 0, \dot{\zeta}_0 = \varepsilon$$

from which it is difficult to get the expected orbit. This is because there is no estimation about the y-axis velocity $\dot{\eta}_0$, which is necessary to generate the vertical orbit. Since the vertical orbit originates from the center manifolds generated from the third pair of pure imaginary eigenvalues and eigenvectors, the initial conditions $\boldsymbol{S}$ that are used to generate the manifolds are applied here for searching for the vertical POs. If this pair of eigenvalues and their corresponding eigenvectors are denoted as $\pm i\beta$ and $u_{Re} \pm i u_{Im}$, respectively, then $\boldsymbol{S}$ can be expressed as [46]

$$\boldsymbol{S} = \boldsymbol{S}_{EP} + 2\varepsilon\,[\cos(\beta\,(t - t_0) + \varphi)\boldsymbol{u}_{Re} - \sin(\beta\,(t - t_0) + \varphi)\boldsymbol{u}_{Im}] \tag{6.12}$$

in which $\boldsymbol{S}_{EP}$ is the state vector of the EP and ε and φ are the arbitrary amplitude and initial phase angle, respectively. At $t = t_0$, the initial state $\boldsymbol{S}_0$ is obtained as

$$\boldsymbol{S}_0 = \boldsymbol{S}_{EP} + 2\varepsilon\,[\cos(\varphi)u_{Re} - \sin(\varphi)u_{Im}] \tag{6.13}$$

For E1 and E2, this family of orbits is symmetric w.r.t. the $x - y$ and $x - z$ planes. Therefore, $\boldsymbol{S}_0$ serves as the input of the DC method to obtain the vertical POs around E1 and E2 at small amplitudes, and similarly the continuation process is applied to generate POs with large amplitudes. However, for the vertical family around E3, due to the asymmetry of the system w.r.t. the $x - z$ plane, the Levenberg-Marquardt method [28], whose application has no requirement on the symmetric property of the system, is applied to find orbits and the continuation process is then applied for large-amplitude orbits. Since E4 is symmetric with E3 with respect to the $x - z$ plane and they share the same eigenvalues (Table 6.3), the vertical families around E4 are also symmetric with the families of POs around E3 and are not shown here.

All the vertical families are illustrated in Fig. 6.8. For E1, E2 and E3, the vertical orbits bend more towards the $x - y$ plane and come close to the surface of the asteroid, as the z-amplitude increases. Similarly, the closer the orbit to the EPs, the stronger in-plane instability it has. However, orbits change from stable to unstable in z-direction when their z-amplitude exceeds 1.04 and 1.08 for E1 and E2, respectively, as indicated by the pentagrams of s_2. The orbits around E3 first reach the largest z-amplitude 1.31, and then bend back down to the asteroid. They always have vertical stability. However, the in-plane motion around E3 change from unstable to stable in the process of the orbits coming back to the body at a z-amplitude of 0.77. Similarly, new families of orbits are expected to be generated at the pentagram points.

In summary, for the above three kinds of POs, the closer the orbit to the EPs, the larger in-plane instability it has. The Lyapunov orbits change from vertically stable to unstable with the increase of orbital amplitudes in the $x - y$ plane. For Halo and Vertical families around E1 and E2, orbits with small z-amplitudes have vertical stability and then become unstable when their z-amplitudes are larger. However, they always have strong in-plane instability. Compared to orbits around E1 and E2, the orbits with similar size around E3 are less unstable. And a small portion of orbits around E3 even have in-plane stability. This is due to the fact that E3 (complex unstable) is less unstable than E1 and E2 which are hyperbolically unstable. Therefore, the general properties of orbital motion in the vicinity of the unstable EPs are characterized. For real missions, orbital control is required to stabilize the unstable orbits.

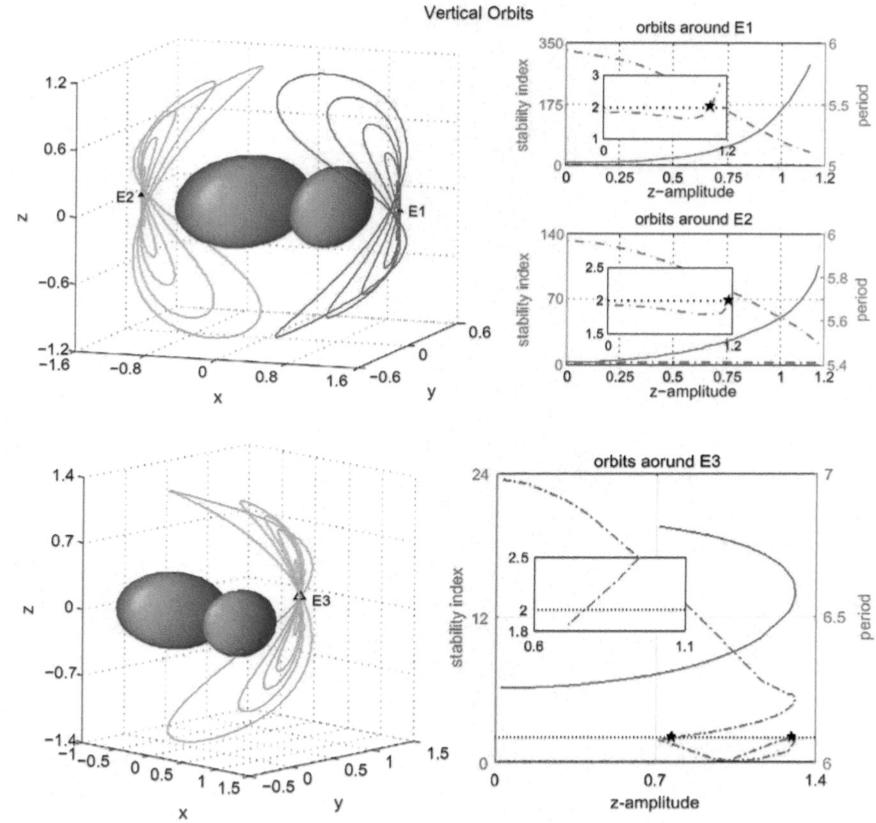

Fig. 6.8 Left: vertical orbits in the vicinity of E1, E2 and E3; Right: their stability index s_1 and s_2 (blue dash-dotted lines) and corresponding periods (orange lines) as a function of the z-amplitude

6.2.4 *Orbital Motion Around the System*

6.2.4.1 Equatorial Orbits

This section focuses on orbits around the entire asteroid system. Given that the rotation rate of the asteroid has value '1' in the normalized system, the initial conditions for an equatorial circular orbit in the rotating frame are expressed as

$$x_{rot} = r, \ y_{rot} = 0, \ \dot{x}_{rot} = 0, \ \dot{y}_{rot} = \pm\sqrt{GM/r} - r = \pm\sqrt{\delta/r} - r$$

where the '+' and '−' signs represent prograde and retrograde orbits in the inertial frame, respectively. The prograde motion in the inertial frame is also prograde in the rotating frame if $\dot{y}_{rot} = \sqrt{\delta/r} - r > 0$ given $\delta > r^3$, and becomes retrograde if

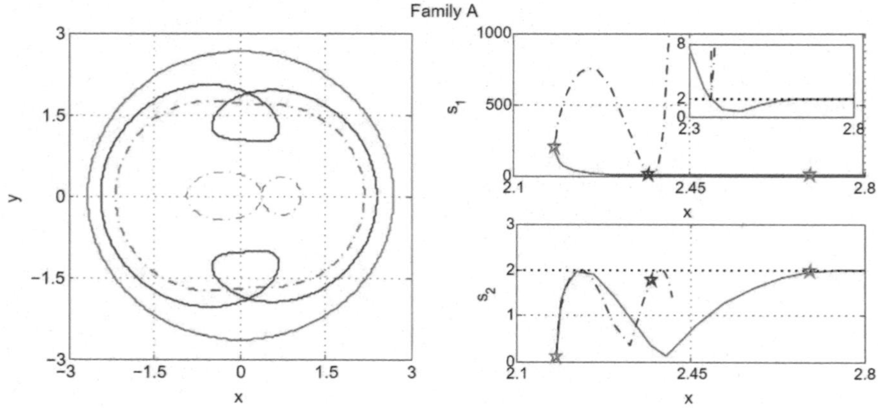

Fig. 6.9 Left: three orbits of family A; Right: the stability indices s_1 and s_2. The pentagrams represent the orbits in the left plot

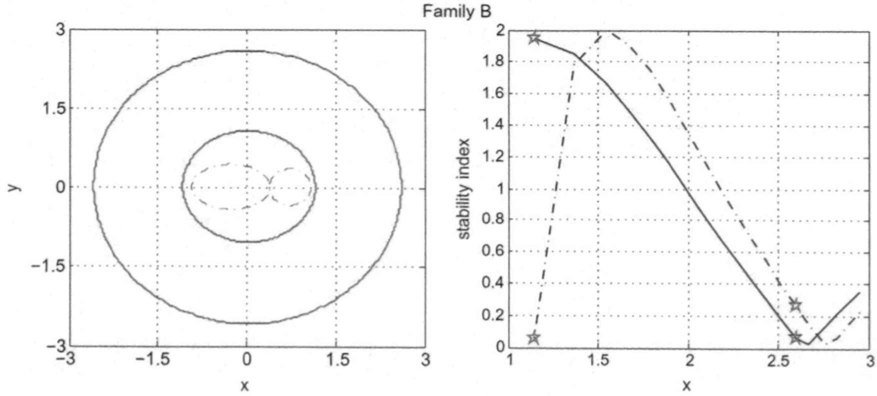

Fig. 6.10 Left: two orbits of family B; Right: the stability indices s_1 (solid line) and s_2 (dash-dotted line). The pentagrams represent orbits in the left plot

$\dot{y}_{rot} = \sqrt{\delta/r} - r < 0$ given $\delta < r^3$. The retrograde motion in the inertial frame is always retrograde in the rotating frame as $\dot{y}_{rot} = -\sqrt{\delta/r} - r < 0$. Therefore, the prograde and retrograde orbits in the inertial frame are denoted as families A and B, respectively. Examples of orbits of families A and B and their stability index are given in Figs. 6.9 and 6.10.

Figure 6.9 illustrates three orbits around the contact binary system of family A, which is found to be all retrograde in the rotating frame. As the orbit comes close to the asteroid, it evolves from circular and stable (red solid) to distorted and unstable (red dash doted). This is also reflected in the stability index plot: the closer the orbit to the asteroid the more unstable it is. In addition, a new family of orbits (blue) is generated at $x \approx 2.4$ as the stability index reaches the value 2, when the bifurcation is expected. The blue orbit in the left plot of Fig. 6.9 belongs to this family, which is characterized by two extra loops in the middle region. Similarly, this family of orbits also becomes highly unstable when approaching the asteroid. However, they all have vertical stability as illustrated by s_3. Therefore, the irregular shape of the asteroid is revealed to have a destabilizing effect on the retrograde A orbits. Two orbits of family B are shown in the left plot of Fig. 6.10; one is the closest orbit of this family around the asteroid with $r \approx 1.2$ and the other one has a radius of 2.6. This family of orbits keeps the circular geometry even when it is in close vicinity of the asteroid. As indicated in the right plot, all these orbits are stable, as their stability indices are not larger than 2. Compared to family A, family B is more robust against the perturbing gravitational field and is preferable for a mission close to the asteroid. This is due to the fact that the relative rotation between family B orbits and the asteroid is generally faster than that of family A, and the perturbation is averaged (or smoothed) to some extent. This analysis is consistent with that of [49]. The influence of the parameter δ on the stability and energy of these retrograde A and B orbits is studied quantitatively in the following section.

6.2.4.2 Effect of Parameter δ on Retrograde A and B Orbits

The above simulations are based on the physical parameters of the system 1996 HW1 with $\delta = 2.1682$. Now the value of δ will be varied from 0.2 to 9, representing a wide range of the rotation rate of the asteroid. The value for μ is kept at the value of 1996 HW1, i.e. 0.2767. Parameter $\delta = 1$ is the situation for fast rotating asteroids, e.g. 2000 EB14 [56]. The case $\delta > 1$ is the more general situation as most contact binary asteroids rotate slowly, where a squeezing force exists between the two components.

Figure 6.11 gives the position-velocity curves and the position-stability index curves of family A orbits at different values of δ, respectively. It is mentioned that

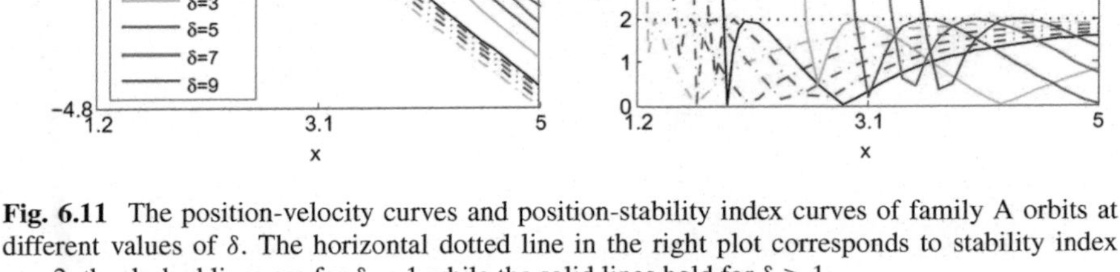

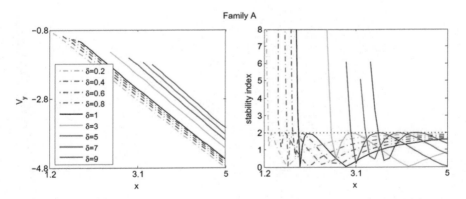

Fig. 6.11 The position-velocity curves and position-stability index curves of family A orbits at different values of δ. The horizontal dotted line in the right plot corresponds to stability index $s = 2$; the dashed lines are for $\delta < 1$ while the solid lines hold for $\delta \geq 1$

all the orbits obtained here do not include the bifurcation families of orbits (which have been shown as the blue orbit in Fig. 6.9), as the main purpose of this section is studying the effect of rotation rate of the asteroid on the orbital motion. It can be seen that all these orbits are retrograde ($V_y < 0$) in the rotating frame. As the asteroid rotates faster (smaller δ) the orbits can be continued to the close vicinity of the asteroid, especially for the case $\delta \leq 1$. For the slowly rotating asteroids (large δ), family A orbits at close distance are not found. For the same x, the faster the rotation of the asteroid, the more initial velocity it requires for the orbital motion in the rotating frame. The stability index is only plotted for s_1, since all orbits have vertical stability. Different with velocity, the stability index does not show a simple change tendency. For all δ, as the orbits come close to the asteroid, after passing a bottom value the stability index touches the line $s = 2$ (indicating bifurcations) and then experiences a low value again. After that it goes up sharply, indicating a fast growth of instability. Furthermore, the faster the rotation of the asteroid, the less the instability of the orbits have and a closer distance to the asteroid stable orbits can be obtained. This is in contrast with the case of the non-collinear EPs (belonging to the 1:1 resonance), which transform from stable to unstable when the rotation rate of the asteroid increases beyond a certain value.

For family B orbits, the position-velocity curves and the position-stability index curves at different values of δ, respectively, are given in Fig. 6.12. In general, for all δ, stable orbits can be obtained in close vicinity of the asteroid, again proving the robustness property of family B orbits against gravitational perturbations. In contrast with family A, for the same x, the faster the rotation of the asteroid (small δ), the less initial velocity it requires for orbital motion in the rotating frame. All these orbits are linearly stable, as their stability index is within the range from 0 to 2. Similar to that of family A, the stability index arrives at the bifurcation point (namely $s = 2$) after experiencing a bottom value close to 0 for the slow rotation cases. For the fast rotation situations, the stability index has a delay to reach the value of 2. In addition, from

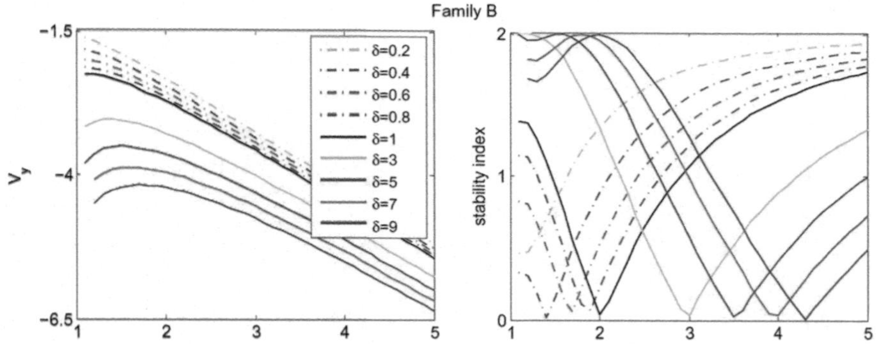

Fig. 6.12 The position-velocity curves and position-stability index curves of retrograde B orbits for different values of δ. The dashed lines are for $\delta < 1$ while the solid lines hold for $\delta \geq 1$

our simulations, it is also found that these orbits are stable at a further distance (e.g. $x = 30$) from the asteroid. In summary, both families of orbits obtained are retrograde in the rotating frame, in which family A has slow retrograde motion and family B has faster one. At the same distance to the asteroid, the orbits are more stable around a fast rotating asteroid than around a slow one. This emphasizes the stabilization effect of the fast rotation of the gravitational field on the equatorial motion, due to the averaging effect of the perturbation. In addition, as the orbit comes close to the asteroid, the dynamics will be subjected to bifurcation earlier for the slowly rotating asteroids. However, when the orbit is extremely close to the asteroid, the fast rotation can no longer diminish the effect of irregular gravity on family A orbits anymore. In general, family B is more stable than family A, and is more suitable for mission orbits.

6.2.4.3 Resonant Orbits

For completeness, the resonant orbits around the entire body are also explored in the rotating frame. Here we define the resonant orbit (with mean angular velocity rate ω_{res}) with commensurability N to the rotation of the asteroid with normalized velocity '1', namely $\omega_{res} = N$. The initial conditions for this $N-$resonance orbit and its orbital period can be derived as

$$x_0 = \sqrt{3\delta}/N^2,\ y_0 = 0,\ z_0 = 0,\ \dot{x}_0 = 0,\ \dot{y}_0 = -N \cdot x_0 - x_0 = -\sqrt{3\delta}N - \sqrt{3\delta}/N^2,$$
$$\dot{z}_0 = \sqrt{\delta/x_0} = \sqrt{3\delta}N,\ T = \begin{cases} 2\pi/N, & N < 1 \\ 2\pi, & N \geq 1 \end{cases}$$

With these initial conditions, the DC method is applied and the polar resonant orbits can be obtained accurately. Examples of orbits at $\delta = 1$ and $\delta = 100$, which respectively represent fast and slow rotation of the asteroid, are given in Figs. 6.13 and 6.14. It can be seen that all these orbits have coverage of the polar region of the asteroid, which is interesting from the perspective of mission design.

For $\delta = 1$, only orbits with $N < 1$ are studied as they physically exist. It is also found that as the orbit comes close to the asteroid, it transits from linearly stable to highly unstable, due to the increasing perturbations from the irregular gravitational field. The three orbits illustrated in Fig. 6.13 are all unstable. For $\delta = 100$, orbits with $N \leq 1$ and with small $N > 1$ (e.g. $N = 1$ and $N = 2$) are obtained, but it is still difficult to identify orbits with large N values. The orbits shown in Fig. 6.14 are also unstable. Different from the equatorial case, for the situation of resonance there is no general conclusion here about the influence of δ on the resonant orbits. The absolute fast rotation of the asteroid is unable to guarantee the stability, as it is already known that the non-collinear EPs that belong to the 1:1 resonance become unstable at the fast rotation case. The commensurability N between the rotation of the asteroid and mean motion of the orbit and δ both play a role on the stability of the resonant orbits.

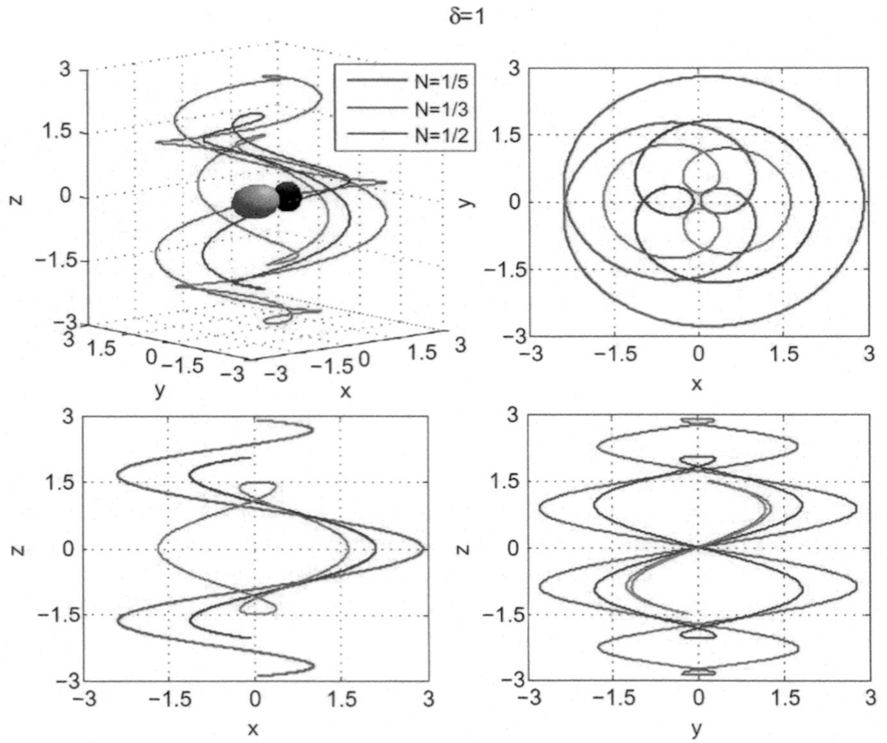

Fig. 6.13 3-dimensional resonant orbits with $N = 1/5, 1/3, 1/2$ at $\delta = 1$ and their projections on the $x - y$, $x - z$ and $y - z$ planes

6.2.5 *Conclusions*

The general dynamical environment around contact binary asteroid systems is explored. Based on the physical parameters of 1996 HW1, four EPs were obtained and their stability was investigated. Families of the Lyapunov, Halo and vertical POs were found in the vicinity of these EPs. It was found that the closer the PO to the EPs, the more unstable it is. The locations of collinear E1 and E2 are sensitive to the change of rotation rate, but not to the system configuration; and they are always unstable. For the non-collinear E3 and E4, system configuration does have a significant influence on their locations. They transit from linear stability to complex instability with increasing rotation rate of the asteroid. The equatorial orbits of families A and B around system 1996 HW1 were addressed. When family A orbits change size from further away from the asteroid to its close vicinity, they transit from stable to highly unstable. This is due to the perturbation from the highly irregular

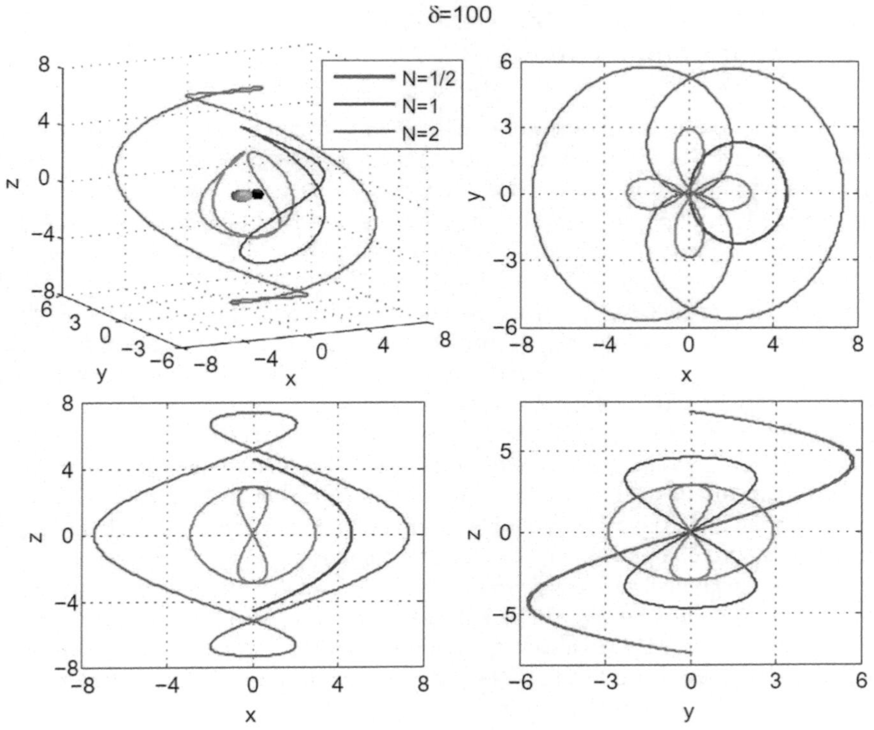

Fig. 6.14 3-dimensional resonant orbits with $N = 1/5, 1/3, 1/2$ at $\delta = 100$ and their projections on the $xy-$, $xz-$and $yz-$planes

gravitational field. Family B orbits were found to remain stable even when they are extremely close to the asteroid and are robust against the gravitational perturbation. Therefore, family B is more preferable for mission orbits. In contrast to the stability of the non-collinear EPs, the fast rotation of the asteroid was proven to have a stabilizing effect on both family A and B orbits, namely the equatorial orbital motion, due to the averaging effect of the irregular gravitation. However, this stabilization cannot diminish the effect of irregular gravity on orbits in the extreme proximity to the asteroid for family A orbits. Finally, resonant orbits were obtained at two rotation rates of the asteroid, i.e. fast case and slow case. They are all unstable but have a good coverage of the polar region of the body. Overall, μ and δ were revealed to have significant influence on the orbital dynamics in the vicinity of the asteroids, and this study contributes to the exploration of the dynamical environment around highly bifurcated bodies.

6.3 Orbital Motion in the Vicinity of the Non-collinear Equilibrium Points

The orbital motion around the non-collinear equilibrium points (EPs) of a contact binary asteroid is investigated. The contact binary asteroid is represented by the combination of an ellipsoid and a sphere. In order to analytically describe the motion around the EPs, different from Sect. 6.1, the gravity field of the ellipsoid is approximated by a spherical harmonic expansion with terms C_{20}, C_{22} and C_{40}, and the sphere by a straightforward point mass model. The non-collinear EPs are linearly stable for asteroids with slow rotation rates, and become unstable as the rotation rate goes up. To study the motion around the stable EPs, a third-order analytical solution is constructed, by the Lindstedt-Poincaré(LP) method. A good agreement is found between this analytical solution and numerical integrations for the motion in the vicinity of the stable EPs. Its accuracy decreases when the orbit goes further away from the EPs and the asteroid rotates faster. For the unstable EPs, the motions around them are unstable as well. Therefore, the linear feedback control law based on low thrust is introduced to stabilize the motion and track the reference trajectory. In addition, more control force is required as any of the injection error, the amplitude of the analytical reference orbit or the rotation rate of the asteroid increases. For small orbits around the EPs, the third-order analytical solution can serve as a good reference trajectory. However, for large amplitude orbits, accurate numerical orbits are to be used as reference. This avoids an extra control force to track the less accurate third-order analytical solution.

6.3.1 Introduction

The highly irregular gravity field induced by a contact asteroid can be modelled with different methods [46]. Outside the circumscribing sphere, a spherical harmonic expansion truncated at arbitrary degree and order can be used. When closer to the body, the polyhedron method of approximation of the shape of a body with triangular faces is more valid [55]. Many studies have been carried out on the dynamical environment around highly bifurcated bodies, e.g. Castalia, Eros and Itokawa [43, 47, 49], mainly focusing on equilibrium points and periodic orbits. The polyhedron method is usually applied for the precise modelling of the gravity field, since it is robust and accurate.

Analytical work is typically to give a more general insight into dynamics, but this is difficult to perform with such a precise model. Therefore, for a general study of this kind of contact binary body, a simplified model of a combination of a sphere and an ellipsoid is applied here, as shown in Fig. 6.2. For this specific configuration, possible formation mechanisms and the relationship between the relative configuration and the rotational angular momentum have been studied in detail in [45]. This simplified model captures the main characteristics of the mass distribution of a contact binary body. Additionally, it breaks one-axial symmetry, which is one step further compared with other totally symmetrical shapes, e.g. two connected spheres [39], two orthogonal segments [1], which were applied for analytical work. In [2], the ellipsoid and the sphere components are separated and serve as the basic model of a binary asteroid pair.

The linear stability of the EPs of an ellipsoid-shaped asteroid has been identified by [44], in which the asteroid is classified as Type I in the case of stable non-collinear EPs and Type II in the case of unstable ones. It is known that the collinear EPs are always unstable, whereas the non-collinear EPs transit from stable to unstable when the rotation rate of the asteroid increases. This is similar to the CRTBP, for which the linear stability of the non-collinear EPs (known as the L_4 and L_5 points) bifurcates at the mass ratio $\mu_c = 0.03852$ (the Routh's critical value). Purely numerical methods have been applied for studying the motion around the EPs for a binary asteroid system [10]. In [20], the first-order solution was described in the form of a series expansion form for motion around the EPs of asteroids with a gravity field from the polyhedron method. However, little work has been done about high-order expansion of the motion around the non-collinear EPs of a contact binary body. This is the focus of the current study, in which the motion is expanded to the third order with the simplified gravity field and the influence of the rotation of the asteroid on the accuracy of the expansion is investigated.

For the stable EPs, a straightforward analytical technique for investigating this motion is the Lindstedt-Poincaré (LP) method. Its basic idea is that the non-linearity of the system alters the frequencies of the linearized system [33]. Extensive use of this technique has been made to obtain periodic solutions of various non-linear systems. One typical application of the LP method in celestial mechanics is the construction of third-order halo orbits around the collinear EPs in the CRTBP [41]. Afterwards, extended work has been carried out by [12, 21, 26], in which analytical solutions in the vicinity of the collinear and non-collinear EPs have been constructed to arbitrary order. Additionally, taking into account the effects of the lunar orbital eccentricity and the solar gravity field in the Earth-Moon system, halo orbits and Lissajous orbits around the trans-lunar libration points have been obtained to the third order [9]. In [13], periodic orbits of the Hill model have been constructed, also by applying the LP method. With the real Earth-Moon model, the EPs of the original CRTBP have been replaced by special quasi-periodic orbits, which are called dynamical substitutes [17]. With the LP method, higher order analytical solutions of the periodic motion in the vicinity of these dynamical substitutes can be constructed [18]. Therefore, in this section, the LP method is also applied to a highly irregular gravity field.

As for the unstable EPs, the orbital motion around them is also unstable. Therefore, a low-thrust control strategy with the optimal linear feedback control law is employed for tracking the nominal orbital motion. The basic idea of this control method is to eliminate the unstable component of the motion and follow the desired trajectory. This method is widely applied for stabilizing unstable non-linear dynamical systems, and is well suited for orbit maintenance around the collinear EPs in the CR3BP [8, 15].

This section is organized as follows. Firstly, a simplified model of a rotating contact binary asteroid is introduced. Secondly, based on the parameters of the system 1996 HW1 which is one of the most bifurcated bodies known to date [31], the location of the non-collinear EPs and their linear stability are investigated at different rotation rates of the asteroid. Thirdly, for the stable EPs, periodic and quasi-periodic orbits are analytically constructed to the third order by means of the LP method. The accuracy of these analytical orbits and the influence of the orbit amplitude and the rotation rate of the asteroid are investigated by numerical integration. Finally, for the unstable EPs, linear feedback control is applied to track the third-order analytical orbit. Again, the influence of the rotation of the asteroid and the orbit amplitude on the propellant consumption is also studied. In addition, for large orbit amplitudes, an accurate orbit numerically derived from the third-order analytical solution is also tracked, to quantify the extra control force consumed by following the less accurate analytical solution.

This study provides a method of studying the dynamics around contact binary bodies, i.e. taking the two components individually from the gravity field point of view. The analytical solutions obtained here can serve as the initial point for solving dynamics in the precise gravity field, once the two components of the body are determined from observations.

6.3.2 Dynamical Model

The geometry of the ellipsoid-sphere configuration is illustrated in Fig. 6.2. The system is assumed to be homogeneous (i.e. the same density for both components) and to rotate uniformly with rotation velocity $\boldsymbol{\omega}$. Different from that of Sect. 6.1, the potential of the ellipsoid component is approximated with the spherical harmonics coefficients truncated at the fourth degree and order. Therefore, the equation of motion in the vicinity of the asteroid is expressed in this frame as

$$\ddot{\boldsymbol{r}} + 2\boldsymbol{\omega} \times \dot{\boldsymbol{r}} + \boldsymbol{\omega} \times (\boldsymbol{\omega} \times \boldsymbol{r}) = \frac{\partial U_{se}}{\partial \boldsymbol{r}} \tag{6.14}$$

where $\boldsymbol{r} = (\tilde{x}, \tilde{y}, \tilde{z})$ is the state vector and

$$\begin{aligned} U_{se} = U_s + U_e &= GM \cdot \left[\frac{\mu}{r_1} + (1-\mu) \cdot U_e(\boldsymbol{r}_2)\right] \\ &= GM \cdot \left\{\frac{\mu}{r_1} + \frac{1-\mu}{r_2} + (1-\mu) \cdot \tilde{C}_{20}\left(\frac{3\tilde{z}^2}{2r_2^5} - \frac{1}{2r_2^3}\right)\right. \\ &\quad \left. + \tilde{C}_{22}\frac{3[(\tilde{x}+\mu)^2 - \tilde{y}^2]}{r_2^5} + \frac{\tilde{C}_{40}}{8}\left(\frac{35\tilde{z}^4}{r_2^9} - \frac{30\tilde{z}^2}{r_2^7} + \frac{3}{r_2^5}\right)\right\} \end{aligned} \tag{6.15}$$

in which U_e is the ellipsoid potential of unit mass in spherical harmonics, $\boldsymbol{r_2} = (\tilde{x}, \tilde{y}, \tilde{z}) - (1-\mu)\boldsymbol{d}$ and $\boldsymbol{r_1} = (\tilde{x}, \tilde{y}, \tilde{z}) + \mu\boldsymbol{d}$ are the vectors from the sphere center at $(1-\mu, 0, 0)$ and ellipsoid center at $(-\mu, 0, 0)$ to the particle at $(\tilde{x}, \tilde{y}, \tilde{z})$, respectively, and $r_1 = |\boldsymbol{r_1}|$, $r_2 = |\boldsymbol{r_2}|$. $\tilde{C}_{20}$, $\tilde{C}_{22}$, $\tilde{C}_{40}$ are the spherical harmonics coefficients of the ellipsoid component. The system can be normalized by taking the length unit as $d = |\boldsymbol{d}|$ and the time unit as $\omega^{-1} = |\boldsymbol{\omega}|^{-1}$. The normalized state vector is written as $\boldsymbol{r_n} = (x, y, z)$. In addition, the normalized spherical harmonic coefficients C_{20}, C_{22} and C_{40} can be derived from the shape of the ellipsoid. They are expressed as

$$\begin{cases} C_{20} = \dfrac{1}{5d^2}\left(c^2 - \dfrac{a^2+b^2}{2}\right) \\ C_{22} = \dfrac{1}{20d^2}\left(a^2 - b^2\right) \\ C_{40} = \dfrac{15}{7}\left(C_{20}^2 + 2C_{22}^2\right) \end{cases} \tag{6.16}$$

in which a, b, c reperesent three semi-axes of the ellipsoid, rather than α, β, γ that are used in the following part as the amplitudes of orbital motion. Since the ellipsoid component is totally symmetric, there are no odd terms. In addition, the higher-order spherical harmonics are ignored as their coefficients are usually very small, e.g. C_{42} and C_{44} are usually one or two orders of magnitude smaller than C_{40}. Their contribution to the results is limited, whereas they introduce significant complexity for solving the dynamics. After normalization, the equations of motion for this dynamical system can be expressed as

$$\begin{cases} \ddot{x} = 2\dot{y} + \dfrac{\partial V}{\partial x} \\ \ddot{y} = -2\dot{x} + \dfrac{\partial V}{\partial y} \\ \ddot{z} = \dfrac{\partial V}{\partial z} \end{cases} \tag{6.17}$$

in which V is the effective potential, written as

$$V = \frac{x^2+y^2}{2} + \delta \cdot \bar{U}_{se}$$

where $\bar{U}_{se}$ is the normalized potential and $\delta = GM/(\omega^2 d^3)$ is the ratio of the gravitational force to the centripetal force that has been defined in Sect. 6.2.2.

6.3.3 Non-collinear EPs and Their Stability

The EPs are defined as the point where all the accelerations are zero. They can be obtained by finding the roots of Eq. (6.17) numerically, and are denoted as (x_e, y_e, z_e). The orbital motion relative to an arbitrary EP can be expressed as

$$\begin{cases} \ddot{\xi} - 2\dot{\eta} = \dfrac{\partial V}{\partial \xi} \\ \ddot{\eta} + 2\dot{\xi} = \dfrac{\partial V}{\partial \eta} \\ \ddot{\varsigma} = \dfrac{\partial V}{\partial \varsigma} \end{cases} \tag{6.18}$$

where $(\xi, \eta, \varsigma)^T = (x, y, z)^T - (x_e, y_e, z_e)^T$ is the position offset w.r.t. the EP. For dynamics around such an asteroid, the location and stability of the EPs are heavily dependent on the gravity field and the rotation rate of the body. Based on the assumption that the asteroid rotates uniformly along the z-direction, the second partial derivatives $V_{xz}, V_{zx}, V_{yz}, V_{zy}$ evaluated at the EPs are all zero. Therefore, the motion in the $x - y$ plane is uncoupled from that along the z-direction for the linearized system which can be expressed as follows

$$\begin{cases} \ddot{\xi} - 2\dot{\eta} = V_{11}\xi + V_{12}\eta \\ \ddot{\eta} + 2\dot{\xi} = V_{21}\xi + V_{22}\eta \\ \ddot{\varsigma} = V_{33}\varsigma \end{cases} \tag{6.19}$$

The partial derivatives V_{ij} evaluated at the non-collinear EP are given in Appendix D. Now, given the parameters of 1996 HW1 (Table 6.1), the EPs are illustrated in Fig. 6.15.

Once the EPs are obtained, their linear stability can be determined from the eigenvalues of the Jacobian matrix [7]. Since the motion in the $x - y$ plane is independent of that along the z-direction, they can be studied separately. The characteristic equation in the $x - y$ plane can be written as

$$\lambda^4 + (4 - V_{11} - V_{22}) \cdot \lambda^2 + V_{11}V_{22} - V_{12}^2 = 0.$$

The $x - y$ plane motion around the EP is stable only when the above equation has two pairs of purely imaginary characteristic roots, denoted by $\pm i\omega_0, \pm i\nu_0$. The following conditions need to be satisfied:

$$\begin{cases} C1 = 4 - V_{11} - V_{22} > 0 \\ C2 = V_{11}V_{22} - V_{12}^2 > 0 \\ C3 = (4 - V_{11} - V_{22})^2 - 4(V_{11}V_{22} - V_{12}^2) > 0 \end{cases} \tag{6.20}$$

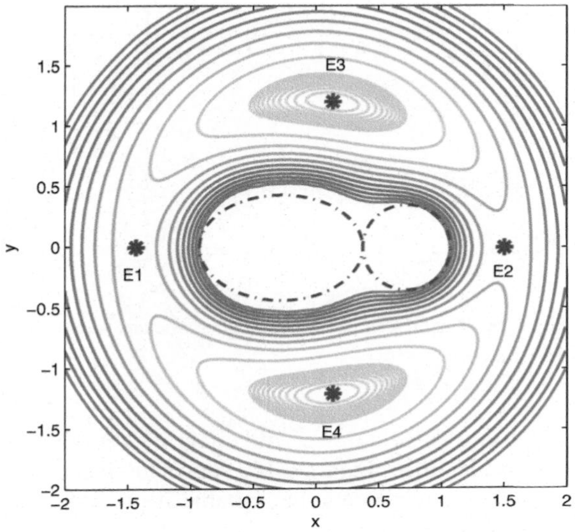

Fig. 6.15 The position of the EPs of system 1996 HW1; E3 and E4 are the non-collinear EPs

For the motion in the z-direction, the characteristic equation is given as

$$\lambda^2 - V_{33} = 0$$

This motion is stable if $V_{33} < 0$ and the characteristic root is represented as $\pm i u_0$. The three frequencies can be obtained as

$$\omega_0 = \sqrt{\frac{C1 - \sqrt{C3}}{2}}, \ v_0 = \sqrt{\frac{C1 + \sqrt{C3}}{2}}, \ u_0 = \sqrt{-V_{33}}$$

It is known that in the CRTBP, the triangular EPs are linearly stable if the mass ratio μ of the system is smaller than $\mu_c = 0.03852$. However, in our problem there is an additional parameter δ (see Sect. 6.3.2) that will have an additional effect on the stability of the non-collinear EPs. For each specific system, there is a critical value δ_c (the bifurcation point for linear stability) which can be determined from numerical simulation. At this point, the system has one pair of pure imaginary eigenvalues of multiplicity two, which means that

$$(4 - V_{11} - V_{22})^2 - 4\left(V_{11}V_{22} - V_{12}^2\right) = 0$$

According to [31], the system 1996 HW1 has $\mu = 0.27673$ and $\delta = 2.1682$. As a result, its non-collinear EPs are unstable. Compared with Fig. 11 from [31], in which a full gravity field is applied, the locations of the EPs we obtained are quite close to theirs and the linear stability of the EPs are the same. This means that the main dynamics of the full gravity field is captured with this simplified model.

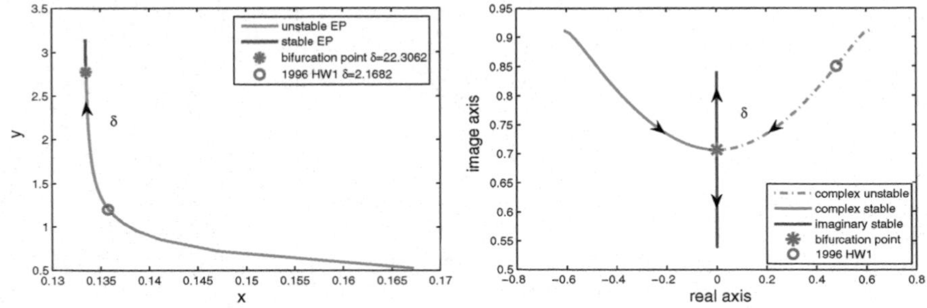

Fig. 6.16 The bifurcation diagram where the arrow represents the evolution of the system when δ increases from 0.3 to 32

Given this value of μ, the effect of δ on the location and stability of the EPs can be identified. Figure 6.16a shows the location of the non-collinear EPs, where the red line represents the unstable EP and the blue line holds for the stable EP. The bifurcation point of this system is found to be $\delta_c = 22.30624$. The bifurcation diagram in Fig.6.16b describes the eigenvalues in the $x - y$ plane.

The red lines represent the two pairs of conjugate complex eigenvalues. The red dotted line in the right half complex plane corresponds to the pair with a positive real component, while the red solid line in the left half plane holds for the pair with negative real parts. As δ increases, the two pairs of complex eigenvalues evolve into pure imaginary eigenvalues as demonstrated by the blue line. Based on the characteristics of the stable and unstable non-collinear EPs, the motion in their vicinity can be studied in detail, which is addressed in the following sections.

6.3.4 Motion Around the Stable Non-collinear EPs

This section focuses on how the LP method is applied in our model, to obtain periodic and quasi-periodic orbits around the stable non-collinear EPs.

6.3.4.1 The Third-Order Analytical Orbits

Following the LP method, the general procedure to construct the analytical solutions from low orders to high orders (to be defined later) is given. It is actually a recursion process which determines the coefficients of each order from the ones of lower orders. The solution of the linear system serves as the starting point. For the stable EPs, Eq. (6.19) have three pairs of pure imaginary roots. Since the out-of-plane motion is uncoupled from the in-plane motion, the general solution of the linearized system Eq. (6.19) can be expressed as

$$\begin{cases} \xi = \alpha \cos(\theta) + \beta \cos(\theta_2) \\ \eta = a_1 \alpha \cos(\theta_1) + b_1 \alpha \sin(\theta_1) + a_2 \beta \cos(\theta_2) + b_2 \beta \sin(\theta_2) \\ \varsigma = \gamma \cos(\theta_3) \end{cases} \tag{6.21}$$

where $\theta_1 = \omega_0 t + \phi_1,\ \theta_2 = v_0 t + \phi_2,\ \theta_3 = u_0 t + \phi_3$, and ω_0, v_0, u_0 are the three frequencies mentioned in Sect. 6.3.3, ϕ_1, ϕ_2, ϕ_3 are the initial phase angles, and α, β, γ are the amplitudes of the corresponding components (It is pointed that they are used to represent the three semi-axes of the ellipsoid in Sect. 6.2.1. But in this section, they represent the amplitudes.). The coefficients in the solution have the following form

$$a_1 = \frac{V_{12} \cdot b_1}{2 \cdot \omega_0},\ b_1 = -\frac{2\omega_0(\omega_0^2 + V_{11})}{4\omega_0^2 + V_{12}^2},\ a_2 = \frac{V_{12} \cdot b_2}{2v_0},\ b_2 = -\frac{2v_0(v_0^2 + V_1 1)}{4v_0^2 + V_{12}^2}$$

When the nonlinear terms are taken into account, the motion can be expanded into trigonometric series of the amplitudes α, β, γ. This is analogous to that of the collinear EPs [21], with the difference that there are three frequencies here. The general form of the high-order solution of Eq. (6.18) can be written as

$$\begin{cases} \xi = \sum\limits_{i,j,k=1}^{\infty} \left[\xi_{ijk}^{lmn} \cos(l\theta_1 + m\theta_2 + n\theta_3) + \bar{\xi}_{ijk}^{lmn} \sin(l\theta_1 + m\theta_2 + n\theta_3)\right] \alpha^i \beta^j \gamma^k \\ \eta = \sum\limits_{i,j,k=1}^{\infty} \left[\eta_{ijk}^{lmn} \cos(l\theta_1 + m\theta_2 + n\theta_3) + \bar{\eta}_{ijk}^{lmn} \sin(l\theta_1 + m\theta_2 + n\theta_3)\right] \alpha^i \beta^j \gamma^k \\ \varsigma = \sum\limits_{i,j,k=1}^{\infty} \left[\varsigma_{ijk}^{lmn} \cos(l\theta_1 + m\theta_2 + n\theta_3) + \bar{\varsigma}_{ijk}^{lmn} \sin(l\theta_1 + m\theta_2 + n\theta_3)\right] \alpha^i \beta^j \gamma^k \end{cases} \tag{6.22}$$

in which $\theta_1 = \omega t + \phi_1, \theta_2 = vt + \phi_2, \theta_3 = ut + \phi_3$ and $\xi_{ijk}^{lmn}, \xi_{ijk}^{lmn}, \eta_{ijk}^{lmn}, \eta_{ijk}^{lmn}, \varsigma_{ijk}^{lmn}, \varsigma_{ijk}^{lmn}$ are the coefficients to be determined at the order of N. Here, the order of the analytical solution is defined as the sum of the three power magnitudes, $N = i + j + k$. The numbers i, j, k are non-negative integers, and l, m, n are the coefficients of the frequencies and also integers. Due to the influence of the nonlinear terms, the frequencies are no longer constant and can be expanded in power series of amplitudes

$$\omega = \omega_0 + \sum_{i,j,k=1}^{\infty} \omega_{ijk} \alpha^i \beta^j \gamma^k,\ v = v_0 + \sum_{i,j,k=1}^{\infty} v_{ijk} \alpha^i \beta^j \gamma^k,\ u = u_0 + \sum_{i,j,k=1}^{\infty} u_{ijk} \alpha^i \beta^j \gamma^k$$

and $\omega_{ijk}, v_{ijk}, u_{ijk}$ are also coefficients to be determined. Due to the characteristics of the LP method and the symmetries of our model, the following conditions are met:

$$\begin{cases} k \text{ is even, } \varsigma_{ijk}^{lmn} = \bar{\varsigma}_{ijk}^{lmn} = 0 \\ k \text{ is odd, } \xi_{ijk}^{lmn} = \bar{\xi}_{ijk}^{lmn} = \eta_{ijk}^{lmn} = \bar{\eta}_{ijk}^{lmn} = 0 \end{cases} \tag{6.23}$$

$|l| \leq i$, $|m| \leq j$, $|n| \leq k$ and l, m, n have the same parity as i, j, k, respectively. Considering the symmetries of the solution, in general $l \leq 0$ is assumed, and $m \leq 0$ is assumed if $l = 0$, $n \leq 0$ is assumed if $l = m = 0$. $\omega_{ijk}, v_{ijk}, u_{ijk}$ are non-zero only if i, j, k are all even numbers.

Suppose we already have the solution up to the order $N - 1$. We substitute it into Eq. (6.18), group all unknown terms of the order N at the left-hand side of the equation and all the known terms of the order N at the right-hand side, then we can solve the unknown terms from the known terms. Generally, the known terms of the order N are only determined by the solution up to the order $N - 1$, so the recursive process is valid. The first-order solution to start the recurrence process is

$$\xi_{100}^{100} = 1,\ \xi_{010}^{010} = 1,\ \eta_{100}^{100} = a_1,\ \bar{\eta}_{100}^{100} = b_1,\ \eta_{010}^{010} = a_2,\ \bar{\eta}_{010}^{010} = b_2,\ \varsigma_{001}^{001} = 1$$

Substitution of the solution up to the order $N - 1$ into the right-hand part of Eq. (6.18) is straightforward. It does not produce any unknown terms of the order N. For the left-hand part of Eq. (6.18), the process is complicated. Take the ξ component as an example; how this process is proceeded is given in Appendix E. Similar results can be derived for $\dot{\eta}, \ddot{\eta}, \dot{\varsigma}, \ddot{\varsigma}$ Thus, equating the coefficients of the order N terms at both sides of the equation will give us the equations for the unknown coefficients, written in matrix form as

$$\begin{bmatrix} A_{11} & A_{12} & A_{13} & A_{14} \\ A_{21} & A_{22} & A_{23} & A_{24} \\ A_{31} & A_{32} & A_{33} & A_{34} \\ A_{41} & A_{42} & A_{43} & A_{44} \end{bmatrix} \begin{bmatrix} \xi_{ijk}^{lmn} \\ \bar{\xi}_{ijk}^{lmn} \\ \eta_{ijk}^{lmn} \\ \bar{\eta}_{ijk}^{lmn} \end{bmatrix} + \begin{bmatrix} \Delta_{\xi} \\ \Delta_{\bar{\xi}} \\ \Delta_{\eta} \\ \Delta_{\bar{\eta}} \end{bmatrix} = \begin{bmatrix} M_{ijk}^{lmn} \\ \bar{M}_{ijk}^{lmn} \\ N_{ijk}^{lmn} \\ \bar{N}_{ijk}^{lmn} \end{bmatrix} \tag{6.24}$$

The components of matrix $\boldsymbol{A}_{4\times4}$ are expressed as

$$\begin{aligned} A_{11} &= -\Lambda^2 - V_{11},\ A_{12} = 0,\ A_{13} = -V_{12},\ A_{14} = -2\Lambda, \\ A_{21} &= A_{12},\ A_{22} = A_{11},\ A_{23} = -A_{14},\ A_{24} = A_{13}, \\ A_{31} &= -V_{21},\ A_{32} = 2\Lambda,\ A_{33} = -\Lambda^2 - V_{22},\ A_{34} = 0, \\ A_{41} &= -A_{32},\ A_{42} = A_{31},\ A_{43} = A_{34},\ A_{44} = A_{33}. \end{aligned}$$

where $\Lambda = l\omega_0 + m v_0 + n u_0$. The Δ_i represents the expressions of frequencies at the order $N - 1$ and solutions at order 1, and can be obtained from Tables 6.1 and 6.2 as

$$\begin{bmatrix} \Delta_{\xi} \\ \Delta_{\bar{\xi}} \\ \Delta_{\eta} \\ \Delta_{\bar{\eta}} \end{bmatrix} = \begin{bmatrix} -\omega_0 - b_1 \\ a_1 \\ -a_1\omega_0 \\ -b_1\omega_0 - 1 \end{bmatrix} \cdot 2\omega_{i-1jk}\delta_{l1}\delta_{m0}\delta_{n0} + \begin{bmatrix} -v_0 - b_2 \\ a_2 \\ -a_2 v_0 \\ -b_2 v_0 - 1 \end{bmatrix} \cdot 2v_{ij-1k}\delta_{l0}\delta_{m1}\delta_{n0}.$$

In addition, for the ς component, the equation is

$$B \cdot \begin{bmatrix} \varsigma_{ijk}^{lmn} \\ \bar{\varsigma}_{ijk}^{lmn} \end{bmatrix} + \begin{bmatrix} \Delta_{\varsigma} \\ \Delta_{\bar{\varsigma}} \end{bmatrix} = \begin{bmatrix} P_{ijk}^{lmn} \\ \bar{P}_{ijk}^{lmn} \end{bmatrix}. \tag{6.25}$$

where $B = -\Lambda^2 - V_{33}$, $\Delta_{\varsigma} = -2u_0 u_{ijk-1}\delta_{l0}\delta_{m0}\delta_{n1}$, $\Delta_{\varsigma} = 0$.

In Eqs. (6.24) and (6.25), M_{ijk}^{lmn}, M_{ijk}^{lmn}, N_{ijk}^{lmn}, N_{ijk}^{lmn} and P_{ijk}^{lmn}, P_{ijk}^{lmn} are the known components of the equations of motion at the corresponding order. Some special cases should be remarked here:

(1) when $n \neq 0$, the determinant of matrix $\boldsymbol{A}$ is non-zero, and Eq. (6.24) can be solved directly.
(2) when $(l, m, n) = (1, 0, 0)$ or $(l, m, n) = (0, 1, 0)$, matrix $\boldsymbol{A}$ is not full rank, then it is assumed that $\xi_{ijk}^{lmn} = \eta_{ijk}^{lmn} = 0$, and ξ_{ijk}^{lmn}, η_{ijk}^{lmn}, ω_{i-1jk}, v_{ij-1k} can be found.
(3) when $(l, m, n) = (0, 0, 0)$, matrix $\boldsymbol{A}$ is also not full rank, then it is assumed that $\xi_{ijk}^{lmn} = \eta_{ijk}^{lmn} = 0$ and ξ_{ijk}^{lmn}, η_{ijk}^{lmn} can be solved.
(4) when $(l, m, n) = (0, 0, 1)$, $\varsigma_{ijk}^{lmn} = \varsigma_{ijk}^{lmn} = 0$ is assumed and u_{ijk-1} is obtained.

Therefore, the Nth-order solution can be obtained by solving the above equations. Neglecting the possible resonances (which brings about small denominator problem), this process can be theoretically carried on to an arbitrary order. The third-order solution is given. The coefficients of the solution are given in Appendix E, based on the parameters of system 1996 HW1 but with $\delta = 30$, in which case the non-collinear EPs are stable (Fig. 6.16). With this solution in series expansion form, periodic and quasi-periodic orbits around the EPs are constructed.

6.3.4.2 Numerical Verification

To test the accuracy of the third-order analytical solution, numerical simulations are performed in the full model (Eq. (6.18)), the results of which are shown in Figs. 6.18, 6.19, 6.20, 6.21, 6.22 and 6.23. They cover all possible motions in the vicinity of a stable non-collinear EP. All units in the figure are dimensionless and the non-collinear EP is located at the origin of the coordinate frame. The amplitudes are chosen such that the size of the orbit remains around the value of 0.1, and the integration time is selected to insure that the general characteristics of the orbital motion are properly reflected. Figures 6.17, 6.18, 6.19 and 6.20 illustrate the periodic orbits and the numerical orbits with one amplitude equal to 0.1, which also means that only one frequency plays a role for each plot. Figures 6.21–6.22 show quasi-periodic orbits

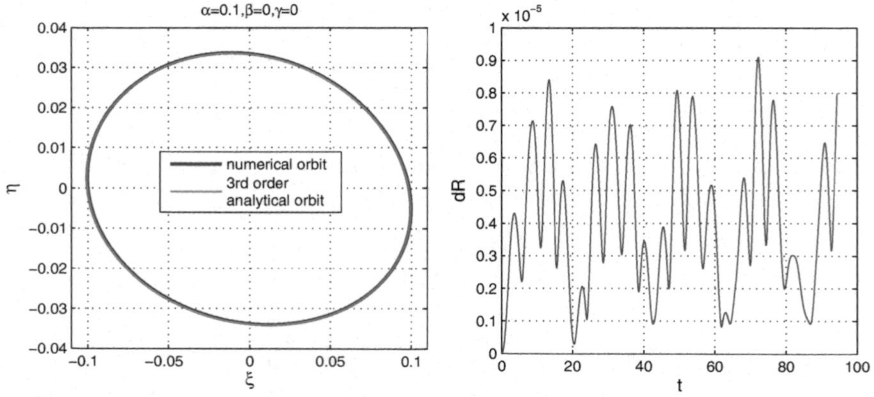

Fig. 6.17 The planar orbit with $\alpha = 0.1$, $\beta = 0$, $\gamma = 0$ (left) and the orbit error (right) for a time duration of 30π

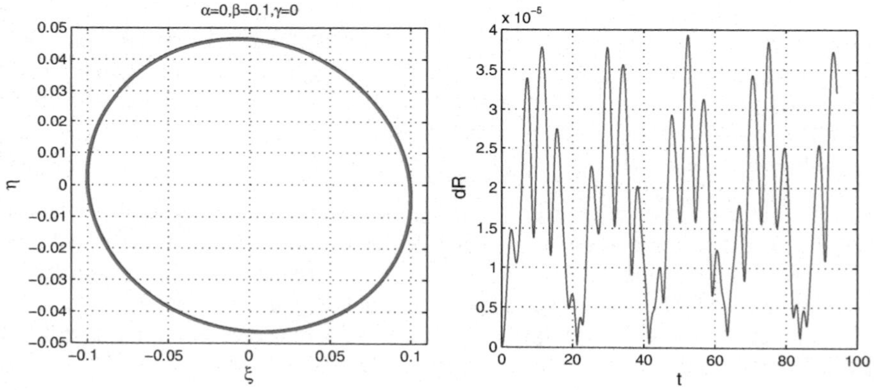

Fig. 6.18 The planar orbit with $\alpha = 0$, $\beta = 0.1$, $\gamma = 0$ (left) and the orbit error (right) for a time duration of 30π

with combinations of two frequencies. Figure 6.21 is the plot of general planar motion with $\gamma = 0$, while Fig. 6.22 is comparable to a Lissajous orbit in the CRTBP. Figure 6.23 illustrates the general three-dimensional quasi-periodic orbit. The error (in the right-hand part of these figures) is defined here as the distance between the analytical and the numerical orbit at each epoch. They are at the magnitude of 10^{-5} in dimensionless units, which indicates that the third-order analytical solution is in very good agreement with its full numerical counterpart.

It is also interesting to investigate how the orbit amplitude and δ influence the error. By numerical simulation, a plot illustrating these relationships is given in Fig. 6.23. It is seen that the maximum error with an integration time of 30π grows when the orbit amplitude increases. This results from the fact that the third-order analytical

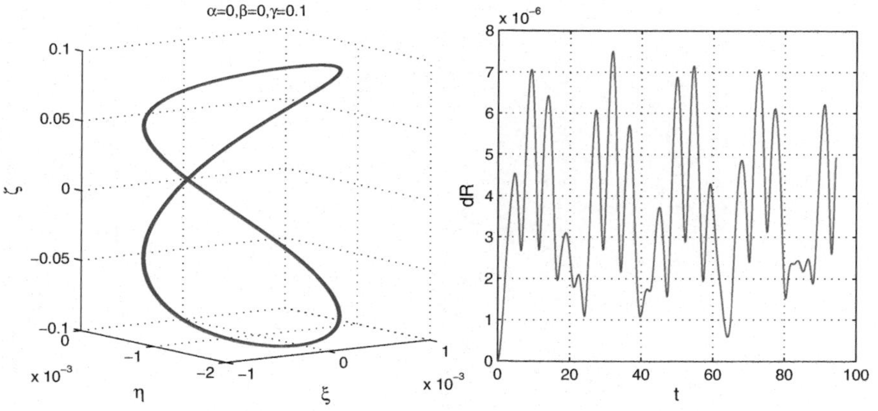

Fig. 6.19 The vertical orbit with $\alpha = 0$, $\beta = 0$, $\gamma = 0.1$ (left) and the orbit error (right) for a time duration of 30π

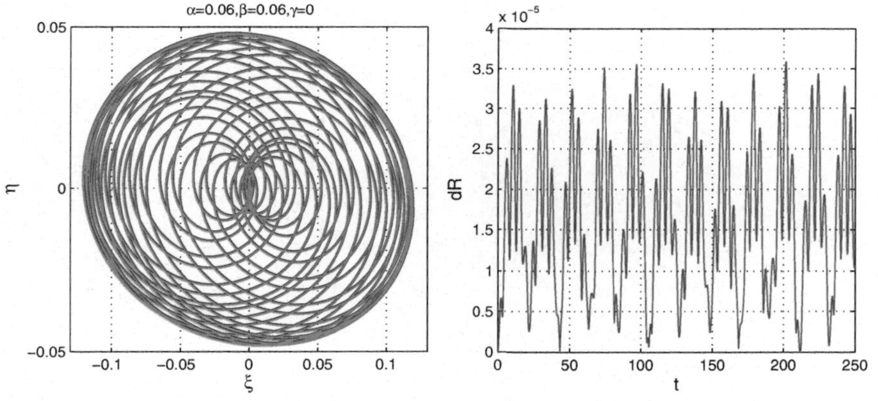

Fig. 6.20 The planar orbit with $\alpha = \beta = 0.06$, $\gamma = 0$ (left) and the orbit error (right) for a time duration of 80π

solution is based on a Taylor-series expansion of the motion in the vicinity of an EP, which is obviously more valid for small deviations. For a given orbit amplitude, the error decreases as δ becomes larger, which implies that the third-order analytical solution is more valid for a system with a slow rotation. In addition, benefiting from all these analyses, the design of orbits can be very straightforward.

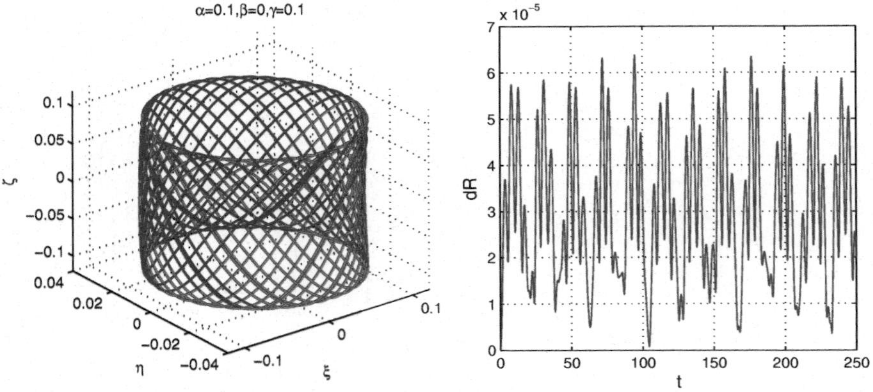

Fig. 6.21 The three-dimensional orbit with $\alpha = \gamma = 0.1$, $\beta = 0$ (left) and the orbit error (right) for a time duration of 80π

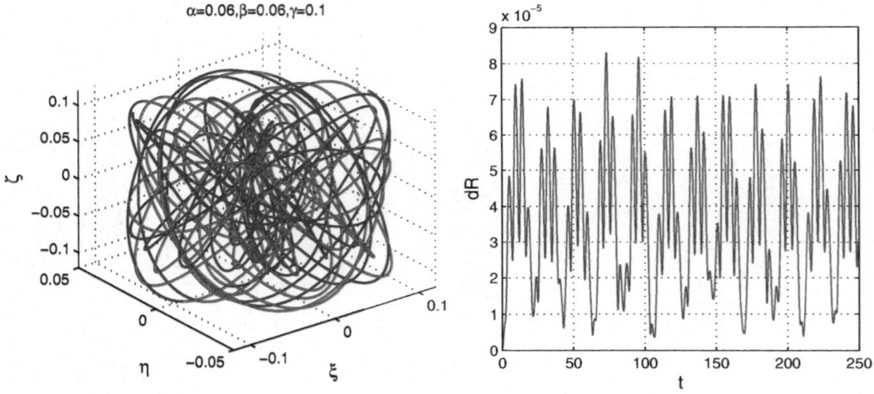

Fig. 6.22 The three-dimensional orbit with $\alpha = \beta = 0.06$, $\gamma = 0.1$ (left) and the orbit error (right) for a time duration of 80π

Fig. 6.23 The influence of orbit amplitude and parameter δ on the error of the third-order analytical solution

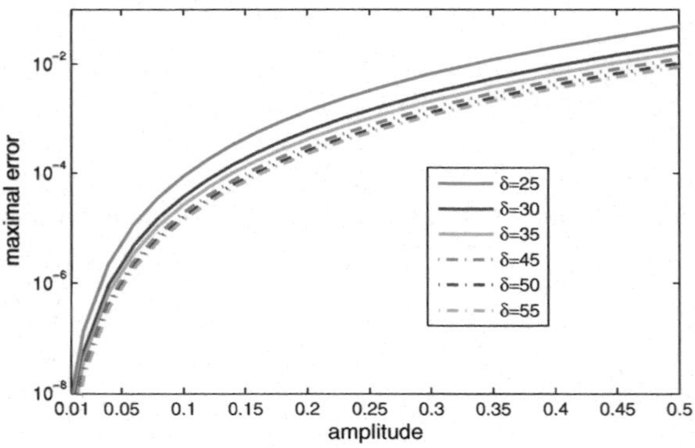

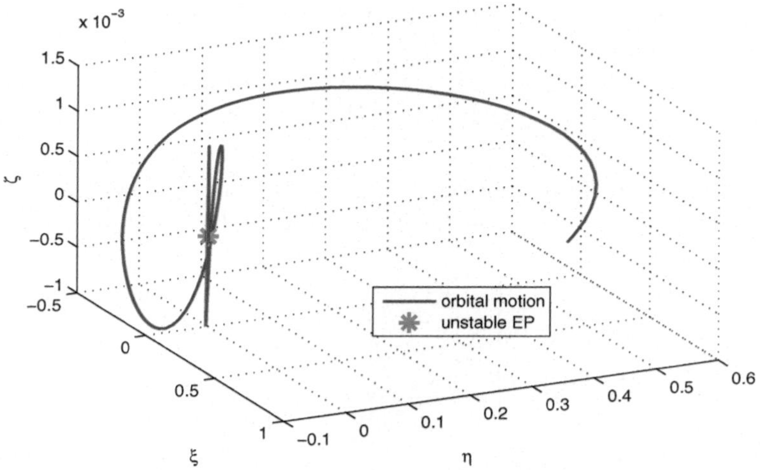

Fig. 6.24 Orbital motion in the vicinity of the unstable non-collinear EP

6.3.5 *Motion Around the Unstable Non-collinear EPs*

For the unstable non-collinear EPs, the situation is quite different. Given a small deviation, the motion around the unstable EP is integrated for a time interval of 5π, as shown in Fig. 6.24. It diverges significantly, especially in the xy-plane, due to the property of instability of the non-collinear EPs.

Therefore, for a spacecraft exploration in this system, low-thrust control is to be introduced to stabilize the motion. A linear feedback control law is employed for stabilizing the unstable non-linear system, which is outlined in the next section.

6.3.5.1 Linear Feedback Control

Firstly, a linearized system with a linear feedback control is written as [51]

$$\dot{\boldsymbol{X}} = \boldsymbol{E} \cdot \boldsymbol{X} + \boldsymbol{F} \cdot \boldsymbol{u} \tag{6.26}$$

where $\boldsymbol{X} = (\Delta\xi, \Delta\eta, \Delta\varsigma, \Delta\dot{\xi}, \Delta\dot{\eta}, \Delta\dot{\varsigma})$ is the deviation vector from the reference trajectory $\boldsymbol{X}_{ref}$ and $\boldsymbol{u} = (u_\xi, u_\eta, u_\varsigma)^T$ is the control vector, $\boldsymbol{E}_{6\times6}$ is the Jacobi matrix calculated along $\boldsymbol{X}_{ref}$, and $\boldsymbol{F}_{6\times3}$ is the control coefficients matrix. The control law is usually defined as $\boldsymbol{u} = -\boldsymbol{K} \cdot \boldsymbol{X}$, in which $\boldsymbol{K}_{3\times6}$ is the gain matrix. Therefore, Eq. (6.26) can be rewritten as $\dot{\boldsymbol{X}} = (\boldsymbol{E} - \boldsymbol{F}\boldsymbol{K}) \cdot \boldsymbol{X}$. By choosing an approximate gain matrix $\boldsymbol{K}$, the new linearized system can be forced to be stable. To meet the tracking accuracy and also the requirement of the allowed control force, a linear quadratic regulator (LQR) is applied to find the optimal control solution and the quadratic cost function is defined as [14]

$$J = \int_0^{\infty} (\boldsymbol{X}^T \boldsymbol{Q} \boldsymbol{X} + \boldsymbol{u}^T \boldsymbol{R} \boldsymbol{u}) dt$$

where $\boldsymbol{Q}$ is a real positive semi-definite matrix and $\boldsymbol{R}$ is a real symmetric positive definite matrix. Both are weighting matrices. They can be defined by the 'Bryson's Rule' [4]

$$\begin{aligned} \boldsymbol{Q} &= diag\left\{\frac{1}{\Delta X_{1max}^2}, \frac{1}{\Delta X_{2max}^2}, \ldots \frac{1}{\Delta X_{6max}^2}\right\}, \\ \boldsymbol{R} &= diag\left\{\frac{1}{\Delta u_{1max}^2}, \frac{1}{\Delta u_{2max}^2}, \frac{1}{\Delta u_{3max}^2}\right\}. \end{aligned} \tag{6.27}$$

where ΔX_{imax}, Δu_{jmax} are the maximum allowable amplitudes of the ith and jth component of $\boldsymbol{X}$ and $\boldsymbol{u}$, respectively. Once $\boldsymbol{Q}$ and $\boldsymbol{R}$ are selected, the control gain matrix $\boldsymbol{K}$ is obtained as

$$\boldsymbol{K} = \boldsymbol{R}^{-1} \boldsymbol{F}^T \boldsymbol{P}$$

where matrix $\boldsymbol{P}_{6\times6}$ is the solution of the following Riccati matrix equation [14]

$$\boldsymbol{E}^T \boldsymbol{P} + \boldsymbol{P} \boldsymbol{E} - \boldsymbol{P} \boldsymbol{F} \boldsymbol{R}^{-1} \boldsymbol{F} \boldsymbol{P} + \boldsymbol{Q} = 0$$

Therefore, the optimal control force is written as

$$\boldsymbol{u} = -\boldsymbol{R}^{-1} \boldsymbol{F}^T \boldsymbol{P} \boldsymbol{X}$$

6.3.5.2 Controlled Motion Tracking the Third-Order Analytical Orbit

Here, the third-order analytical solution will serve as the reference trajectory. As shown in Sect. 6.2.3, the unstable EP has only one pure imaginary eigenvalue in the z-direction. Therefore, the reference trajectory can be given by the third-order analytical solution with $\alpha = \beta = 0$ and γ with a chosen value. Three cases will be studied for the control problem: (1) the spacecraft is exactly put on the reference orbit, in which situation there is no injection error; (2) an initial error, e.g. $[10^{-4}, 10^{-4}, 0, 10^{-4}, -10^{-4}, 0]$, is added to the initial state of the spacecraft; (3) the spacecraft is put on the stable manifold of the EP. For the last case, the local stable manifold of the EP is given as

$$\begin{cases} \xi(t) = e^{(-M\cdot t)} \cdot \big(\gamma_1 \cdot \cos(N\cdot t) + \gamma_2 \cdot \sin(N\cdot t)\big) \\ \eta(t) = \sigma \cdot e^{(-M\cdot t)} \cdot \big(\gamma_3 \cdot \cos(N\cdot t) + \gamma_4 \cdot \sin(N\cdot t)\big) \\ \varsigma(t) = \varsigma_3(t) \end{cases} \tag{6.28}$$

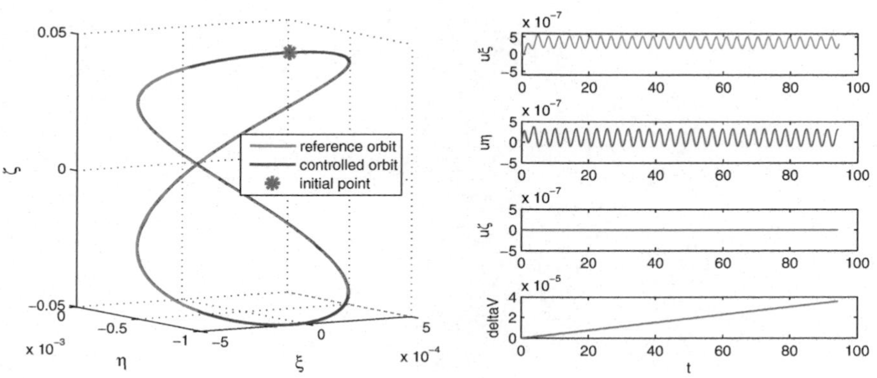

Fig. 6.25 The reference and controlled orbits, and the control accelerations for case 1

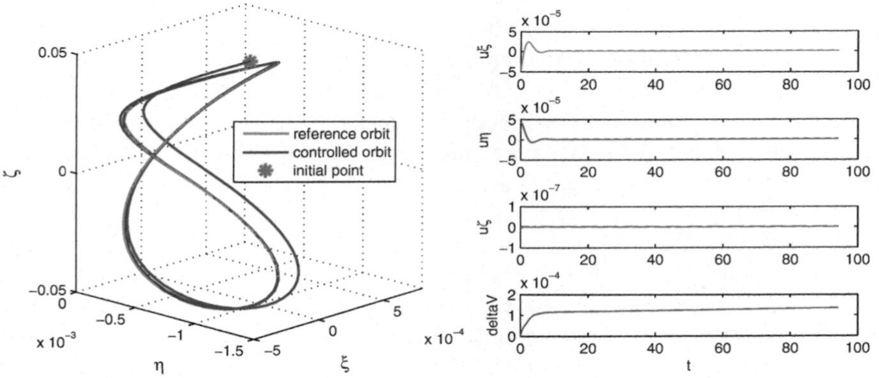

Fig. 6.26 The reference and controlled orbits, and the control accelerations for case 2

in which

$$\begin{cases} \sigma = 1/(M^2 - 4M \cdot V_{12} + V_{12}^2 + 4N^2)\ , \\ \gamma_3 = -\ (2M^3 \cdot \gamma_1 - 2M^2N \cdot \gamma_2 + 2MN^2 \cdot \gamma_1 + V_{12}V_{11} \cdot \gamma_1 + V_{12} \cdot \gamma_1 \cdot N^2 \\ \qquad - 2V_{11} \cdot \gamma_2 \cdot N - 2 \cdot \gamma_2 \cdot N^3 - M^2 V_{12} \cdot \gamma_1 + 2MN \cdot V_{12} \cdot \gamma_2 - 2M \cdot V_{11} \cdot \gamma_1), \\ \gamma_4 = -\ (2M^3 \cdot \gamma_2 + 2M^2N \cdot \gamma_1 + 2MN^2 \cdot \gamma_2 + V_{12}V_{11} \cdot \gamma_2 - V_{12} \cdot \gamma_2 \cdot M^2 \\ \qquad - 2M \cdot V_{11} \cdot \gamma_2 + 2 \cdot \gamma_1 \cdot N^3 + N^2 V_{12} \cdot \gamma_2 - 2MN \cdot V_{12} \cdot \gamma_1 + 2N \cdot V_{11} \cdot \gamma_1). \end{cases} \tag{6.29}$$

where $\gamma_1, \gamma_2, \gamma_3, \gamma_4$ are the amplitudes and M, N are the absolute values of the real and imaginary components of the complex eigenvalues, $\varsigma_3(t)$ is the third-order analytical solution in the z-direction. Based on the parameters of system 1996 HW1 with its value for $\delta = 2.1682$, the reference orbit and controlled motion as well as the control accelerations are given in Figs. 6.25, 6.26 and 6.27.

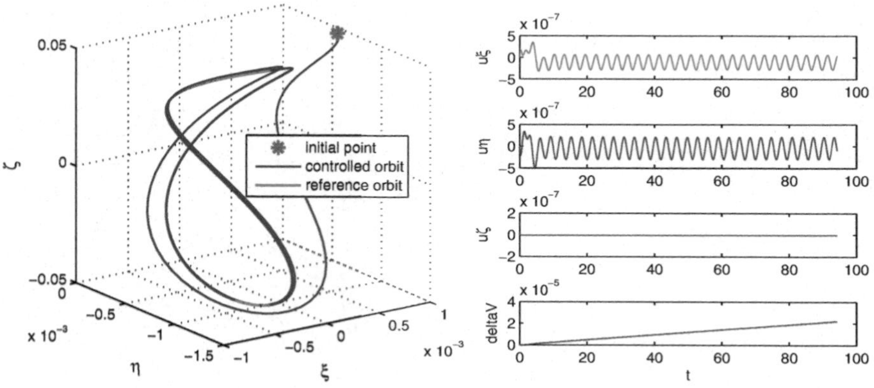

Fig. 6.27 The reference and controlled orbits, and the control accelerations for case 3

For a time interval of 15 revolutions, the propellant needed for the above three cases in dimensionless units are 3.56×10^{-5}, 1.34×10^{-4} and 2.22×10^{-5}, respectively. Going back to dimensional units, it is appealing to notice that about 10^{-5} m/s propellant is required for the spacecraft to stay on this orbit for about 34.4 days, provided no initial offset is present, i.e. case (1). If the spacecraft is put on the local stable manifold of the EP, the consumption is slightly less than case (1) and significantly less than the case when there is a random error in the initial state, i.e. case (2). More generally, we find that the propellant consumption for cases (1) and (3) is close to each other and always smaller than that of case (2), for orbit amplitudes ranging from 0.01 to 0.1. From simulations, it is observed that for case (2) the magnitude of the control delta V is closely related to the magnitude of the initial error. However, since this study is more interested in the influence of the orbit amplitude and parameter δ on the control effort, the exact value of the initial error in case (2) is not so important. Therefore, to avoid the calculation of manifolds in each simulation and also to eliminate the influence of the initial error, case (1) is selected for further study.

As shown in Fig. 6.28, more propellant is needed when the orbit amplitude increases and when the rotation of the asteroid goes faster (smaller δ). Two reasons are responsible for this. One is obviously that more control force is required due to the larger instability of the orbit induced by the faster rotation of the asteroid. The other is due to the fact that the larger the amplitude of the orbit the larger the error of the analytical solution, which adds an extra artificial control effort to follow this solution. This is further studied in the following section.

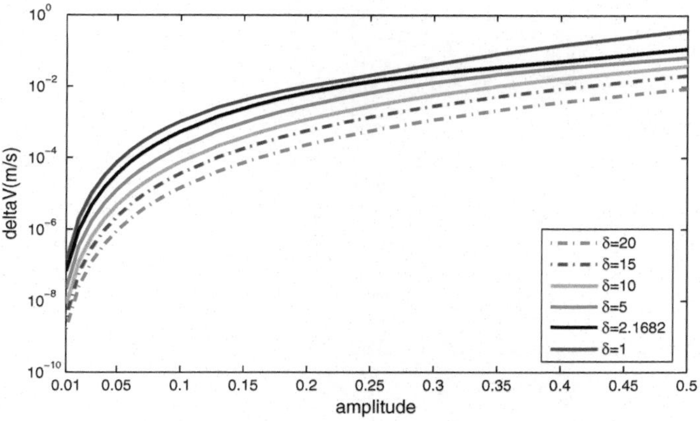

Fig. 6.28 The influence of orbit amplitude and parameter δ on propellant consumption

6.3.5.3 Controlled Motion Tracking the Numerical Orbit

Firstly, exact periodic orbits (POs) are obtained by numerically modifying the third-order analytical solution. The differential correction [24, 42] and the Levenberg-Marquardt method [34] are methods commonly used for numerical corrections. For this simulation, the latter one is applied. The difference between the third-order analytical orbits and the accurate POs is illustrated in Fig. 6.29a. It is less than 10^{-5} when the amplitude of the orbit is smaller than 0.1, and goes up to 10^{-2} when the orbital amplitude enlarges to 0.5. When the asteroid rotates fast, the error becomes larger for orbits at the same amplitude. This is consistent with what was obtained for the situation of stable motion (Fig. 6.23). In addition, the instability of the numerical POs is clarified by the stability index, which is defined as the sum of the norm of each pair of eigenvalues of the monodromy matrix of the PO [7]. If larger than 6, the orbit is unstable. The POs are highly unstable for the rapid rotation case, as illustrated in Fig. 6.29b.

To have a detailed idea of how the error of the analytical PO affects the control effort, a comparison is made between the control force required to track the third-order analytical solution and to track the numerical PO. For the case of system 1996 HW1 and orbit amplitudes ranging from 0.1 to 0.5, Fig. 6.30a illustrates that almost more than half of the propellant is introduced by the error of the analytical reference orbit for a time interval around 4π (about two periods for each orbit). However, the percentage of the control force from tracking the analytical orbit will reduce gradually for longer time intervals, as more control effort is allocated for stabilizing the orbit, as shown in Fig. 6.30b. Actually, this simulation emphasizes that an accurate numerical reference orbit should be used for large amplitude motions to reduce the control force.

Nonetheless, the third-order analytical solution is of important value. It approximates the small-amplitude motion quite well, and also provides a very good initial

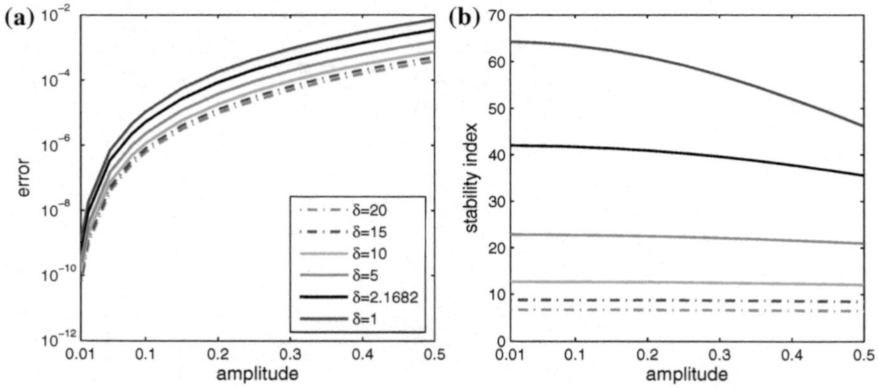

Fig. 6.29 The errors of the analytical orbits (left) and the stability index of the accurate numerical orbits (right) for different orbit amplitudes

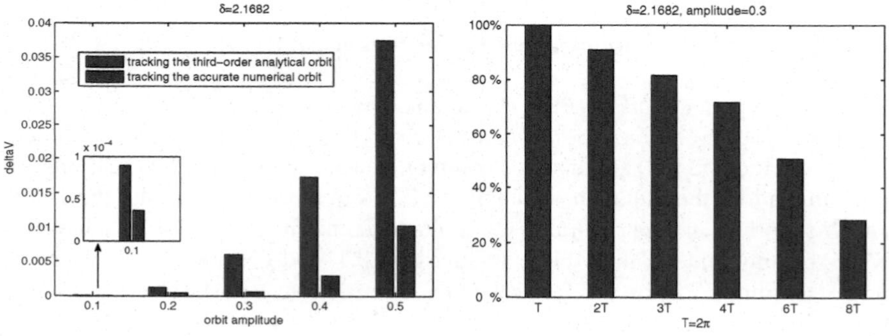

Fig. 6.30 The comparison of delta V between tracking the third-order analytical solution and the numerical orbit for a time duration around 4π (left); the percentage of propellant consumption for compensating the error of the analytical orbit (right) for different time intervals

guess for the numerical search for the accurate numerical orbit when the orbital amplitude is large.

6.3.6 Conclusions

The orbital motion around the non-collinear EPs of a contact binary asteroid has been studied. With the asteroid represented by a combination of an ellipsoid and a sphere and sizes and mass taken from the system 1996 HW1, the locations and the linear stability of the EPs of the system are obtained first for different values of δ.

For motions around the stable EPs, third-order analytical solutions have been constructed, by means of the LP method. Compared with numerical integrations, they are found to be a good approximation for the motions in the vicinity of the stable EPs. However, the accuracy of this third-order analytical solution decreases when the orbital amplitude grows larger and the asteroid rotates faster, in which cases higher order solutions are probably more valid.

In the case of unstable EPs, a low-thrust control strategy based on the linear feedback control law has been employed to follow the third-order reference trajectory. Very little control effort is needed for all three simulated cases. For the spacecraft located on the local stable manifold of the EP, the propellant consumption is found to be approximately the same as the case without initial error. On the other hand, injection error not on the stable manifolds definitely brings about extra control effort. In addition, more propellant is required when the orbit amplitude increases and when the rotation of the asteroid goes faster, which is partially due to the increasing error of the analytical reference orbit. Thus, for large amplitude motion, the accurate orbit obtained from numerically modifying the analytical solution is used as the reference orbit. Consequently, the propellant consumption reduces compared to that tracking the less accurate third-order analytical solution. In conclusion, to save control effort, an accurate numerical orbit is recommended for a mission with large amplitude motion in a highly perturbed environment.

This study can also be extended to binary asteroids, planets and even star systems in which the primary body has an irregular gravity field.

References

1. P. Bartczak, S. Breiter, Double material segment as the model of irregular bodies. Celest. Mech. Dyn. Astron. **86**, 131–141 (2003)
2. J. Bellerose, D.J. Scheeres, The restricted full three body problem: applications to binary asteroid exploration. Ph.D. Dissertation, University of Michigan, 2008
3. R. Broucke, Stability of periodic orbits in the elliptic, restricted three-body problem. AIAA J. **7**(6), 1003–1009 (1969)
4. A.E. Bryson, *Applied Optimal Control: Optimization, Estimation and Control* (CRC Press, Boca Raton, 1975)
5. M. Ceccaroni, J. Biggs, Analytic perturbative theories in highly inhomogeneous gravitational fields. Icarus **224**(1), 74–85 (2013)
6. B. Chauvineau, P. Farinella, F. Mignard, Planar orbits about a triaxial body: application to asteroidal satellites. Icarus **105**(2), 370–384 (1993)
7. C.C. Chicone, *Ordinary Differential Equations with Applications* (Springer, Berlin, 1999)
8. R.W. Farquhar, The control and use of libration-point satellites. National Aeronautics and Space Administration, 1970
9. R.W. Farquhar, A.A. Kamel, Quasi-periodic orbits about the translunar libration point. Celest. Mech. **7**, 458–473 (1973)
10. F. Gabern, W. Koon, J. Marsden, D.J. Scheeres, Binary asteroid observation orbits from a global dynamical perspective. SIAM J. Appl. Dyn. Syst. **5**, 252–279 (2006)
11. D. German, A.L. Friedlander, A simulation of orbits around asteroids using potential field modelling, *Spaceflight Mechanics* (1991)
12. G. Gómez, *Dynamics and Mission Design Near Libration Points: Advanced Methods for Triangular Points*, vol. 4 (World Scientific, Singapore, 2001)

13. G. Gómez, M. Marcote, High-order analytical solutions of Hill's equations. Celest. Mech. Dyn. Astron. **94**, 197–211 (2006)
14. M. Gopal, *Modern Control System Theory* (New Age International, 1993)
15. P. Gurfil, D. Meltzer, Station keeping on unstable orbits: generalization to the elliptic restricted three-body problem. J. Astronaut. Sci. **54**, 29–51 (2006)
16. J.K. Harmon, M.C. Nolan, E.S. Howell, J.D. Giorgini, P.A. Taylor, Radar observations of comet 103P/Hartley 2. Astrophys. J. Lett. **734**, L2 (2011)
17. X. Hou, L. Liu, On quasi-periodic motions around the triangular libration points of the real Earth-Moon system. Celest. Mech. Dyn. Astron. **108**, 301–313 (2010)
18. X. Hou, L. Liu, On quasi-periodic motions around the collinear libration points of the real Earth-Moon system. Celest. Mech. Dyn. Astron. **110**, 71–98 (2011)
19. K.C. Howell, Three-dimensional periodic 'halo' orbits. Celest. Mech. **32**(1), 53–71 (1984)
20. Y. Jiang, H. Baoyin, J. Li, H. Li, Orbits and manifolds near the equilibrium points around a rotating asteroid. Astrophys. Space Sci. **349**, 83–106 (1999)
21. A. Jorba, J. Masdemont, Dynamics in the center manifold of the collinear points of the restricted three body problem. Phys. D Nonlinear Phenom. **132**, 189–213 (1999)
22. M. Lara, Repeat ground track orbits of the Earth tesseral problem as bifurcations of the equatorial family of periodic orbits. Celest. Mech. Dyn. Astron. **86**(2), 143–162 (2003)
23. M. Lara, A. Elipe, Periodic orbits around geostationary positions. Celest. Mech. Dyn. Astron. **82**(3), 285–299 (2002)
24. M. Lara, R. Russell, Computation of a science orbit about Europa. J. Guid. Control Dyn. **30**, 259–263 (2007)
25. M. Lara, D.J. Scheeres, Stability bounds for three-dimensional motion close to asteroids. J. Astronaut. Sci. **50**(4), 389–409 (2002)
26. H. Lei, B. Xu, High-order analytical solutions around triangular libration points in the circular restricted three-body problem. Mon. Not. R. Astron. Soc. **434**, 1376–1386 (2013)
27. X. Liu, H. Baoyin, X. Ma, Equilibria, periodic orbits around equilibria, and heteroclinic connections in the gravity field of a rotating homogeneous cube. Astrophys. Space Sci. **333**(2), 409–418 (2011)
28. M.I. Lourakis, A brief description of the Levenberg-Marquardt algorithm implemented by levmar. Found. Res. Technol. **4**, 1–6 (2005)
29. W.D. MacMillan, *The Theory of the Potential (Chapter 2)* (Dover, New York, 1958)
30. C. Magri, S.J. Ostro, D.J. Scheeres et al., Radar observations and a physical model of Asteroid 1580 Betulia. Icarus **186**, 152–177 (2007)
31. C. Magri, E.S. Howell, M.C. Nolan et al., Radar and photometric observations and shape modeling of contact binary near-Earth Asteroid (8567) 1996 HW1. Icarus **214**, 210–227 (2011)
32. J.D. Meiss, *Differential Dynamical Systems* (SIAM, 2007)
33. R.E. Mickens. An introduction to non-linear oscillations. CUP Archive,1981
34. J.J. More, The Levenberg-Marquardt algorithm: implementation and theory, *Numerical Analysis* (Springer, Berlin, 1978), pp. 105–116
35. F.R. Moulton, D. Buchanan, T. Buck, F.L. Griffin, W.R. Longley, W.D. MacMillan, *Periodic Orbits (Chapter 5)* (Carnegie Institution of Washington, Washington, 1920)
36. P.M. Muller, W.L. Sjogren, Mascons: lunar mass concentrations. Science **161**(3842), 680–684 (1968)
37. H. Osinga, J.G.-V.B. Krauskopf, *Numerical Continuation Methods for Dynamical Systems (Chapter 1)* (Springer, Berlin, 2007)
38. W.H. Press, *Numerical Recipes 3rd Edition: The Art of Scientific Computing (Chapter 6)* (Cambridge University Press, Cambridge, 2007)
39. T. Prieto-Llanos, M.A. Gómez-Tierno, Station keeping at libration points of natural elongated bodies. J. Guid. Control Dyn. **17**, 787–794 (1994)
40. A. Riaguas, A. Elipe, M. Lara, Periodic orbits around a massive straight segment. Celest. Mech. Dyn. Astron. **73**(1–4), 169–178 (1999)
41. D.L. Richardson, Analytic construction of periodic orbits about the collinear points. Celest. Mech. **22**, 241–253 (1980)

42. R.P. Russell, M. Lara, Long-life time lunar repeat ground track orbits. J. Guid. Control Dyn. **30**, 982–993 (2007)
43. D. Scheeres, S. Broschart, S. Ostro, L. Benner, The dynamical environment about Asteroid 25143 Itokawa: target of the Hayabusa Mission, in *AIAA/AAS Astrodynamics Specialist Conference and Exhibit* (2004)
44. D.J. Scheeres, Dynamics about uniformly rotating triaxial ellipsoids: applications to asteroids. Icarus **110**, 225–238 (1994)
45. D.J. Scheeres, Rotational fission of contact binary asteroids. Icarus **189**, 370–385 (2007)
46. D.J. Scheeres, *Orbital Motion in Strongly Perturbed Environments* (Springer, Berlin, 2012)
47. D.J. Scheeres, S.J. Ostro, R. Hudson, R.A. Werner, Orbits close to asteroid 4769 Castalia. Icarus **121**, 67–87 (1996)
48. D.J. Scheeres, S.J. Ostro, R. Hudson, E.M. DeJong, S. Suzuki, Dynamics of orbits close to asteroid 4179 Toutatis. Icarus **132**(1), 53–79 (1998)
49. D. Scheeres, B. Williams, J. Miller, Evaluation of the dynamic environment of an asteroid: applications to 433 Eros. J. Guid. Control Dyn. **23**, 466–475 (2000)
50. H. Sierks, C. Barbieri, P.L. Lamy, R. Rodrigo, D. Koschny, H. Rickman, F. Angrilli, On the nucleus structure and activity of comet 67P/Churyumov-Gerasimenko. Science **347**(6220), aaa1044 (2015)
51. J.-J.E. Slotine, W. Li, *Applied Nonlinear Control* (Prentice-Hall, Englewood Cliffs, NJ, 1991)
52. V. Szebehely, *Theory of Orbits: The Restricted Three Body Problem (Chapter 5)* (Academic Press, San Diego, 1967)
53. O. Vasilkova, Three-dimensional periodic motion in the vicinity of the equilibrium points of an asteroid. Astron. Astrophys. **430**(2), 713–723 (2005)
54. R.A. Werner, The gravitational potential of a homogeneous polyhedron or don't cut corners. Celest. Mech. Dyn. Astron. **59**(3), 253–278 (1994)
55. R.A. Werner, D.J. Scheeres, Exterior gravitation of a polyhedron derived and compared with harmonic and mascon gravitation representations of asteroid 4769 Castalia. Celest. Mech. Dyn. Astron. **65**, 313–344 (1997)
56. R.J. Whiteley, D.J. Tholen, C.W. Hergenrother, Lightcurve analysis of four new monolithic fast-rotating asteroids. Icarus **157**(1), 139–154 (2002)

Appendix A
The Primary Zonal and Tesseral Terms Contributing to the 1:1 Resonance

The primary zonal and tesseral terms contributing to the 1:1 resonance

n	2	2	3	3	4	4	4
m	0	2	1	3	0	2	4
p	1	0	1	0	2	1	0
q	0	0	0	0	0	0	0
Θ_{nmpq}	0	2σ	σ	3σ	0	2σ	4σ

The expressions of $\mathscr{H}_1$ and $\mathscr{H}_2$ are given as

$$\begin{aligned}
\mathscr{H}_1 = &-\frac{\mu^4 R^2}{L^6}[F_{210}G_{20-1}(C_{21}\cos(\sigma + g) + S_{21}\sin(\sigma + g)) \\
&+ F_{211}G_{211}(C_{21}\cos(\sigma - g) + S_{21}\sin 2(\sigma - g))] \\
&- \frac{\mu^5 R^3}{L^8}[F_{301}G_{31-1}(C_{30}\cos(g) + S_{30}\sin(g)) \\
&+ F_{302}G_{321}(C_{30}\cos(g) - S_{30}\sin(g))] \\
&- \frac{\mu^6 R^4}{L^{10}}[F_{411}G_{41-1}(C_{41}\cos(\sigma + g) + S_{41}\sin(\sigma + g)) \\
&+ F_{412}G_{421}(C_{41}\cos(\sigma - g) + S_{41}\sin(\sigma - g)) \\
&+ F_{430}G_{40-1}(C_{43}\cos(3\sigma + g) + S_{43}\sin(3\sigma + g)) \\
&+ F_{431}G_{411}(C_{43}\cos(3\sigma - g) + S_{43}\sin(3\sigma - g))]
\end{aligned} \tag{A.1}$$

J. Yuan et al., *Low Energy Flight: Orbital Dynamics and Mission Trajectory Design*, https://doi.org/10.1007/978-981-13-6130-2

$$
\begin{aligned}
\mathscr{H}_2 = &-\frac{\mu^4 R^2}{L^6}[F_{221}G_{212}(C_{22}\cos(2\sigma - 2g) + S_{22}\sin(2\sigma - 2g))\\
&-\frac{\mu^5 R^3}{L^8}[F_{310}G_{30-2}(C_{31}\cos(\sigma + 2g) + S_{31}\sin(\sigma + 2g))\\
&+ F_{312}G_{322}(C_{31}\cos(\sigma - 2g) + S_{31}\sin(\sigma - 2g))\\
&+ F_{331}G_{312}(C_{33}\cos(3\sigma - 2g) + S_{33}\sin(3\sigma - 2g))]\\
&-\frac{\mu^6 R^4}{L^{10}}[F_{401}G_{41-2}(C_{40}\cos(2g) + S_{40}\sin(2g))\\
&+ F_{422}G_{422}(C_{42}\cos(2\sigma - 2g) + S_{42}\sin(2\sigma - 2g))]
\end{aligned}
\tag{A.2}
$$

Appendix B
The Un-normalized Spherical Harmonic Coefficients to Degree and Order 4

The tables below contain the values for the un-normalized spherical harmonic coefficients to degree and order 4 for Vesta derived from [1], 1996 HW1 from [2] and Betulia derived from [3]. Although there is an update of the gravitational field of Vesta in [4], the full 4 × 4 spherical harmonics are not directly available from it. In addition, the difference between the two is quite small. (The References here are from Chap. 5.)

Vesta					
C_{20}	-6.872555×10^{-2}	S_{31}	1.825409×10^{-4}	S_{41}	-1.347130×10^{-4}
C_{21}	0	C_{32}	-3.162892×10^{-4}	C_{42}	-3.152856×10^{-5}
S_{21}	0	S_{32}	5.943231×10^{-5}	S_{42}	6.551679×10^{-5}
C_{22}	3.079667×10^{-3}	C_{33}	2.565757×10^{-5}	C_{43}	-3.113571×10^{-5}
S_{22}	0	S_{33}	7.264998×10^{-5}	S_{43}	-2.689264×10^{-6}
C_{30}	6.286305×10^{-3}	C_{40}	9.6×10^{-3}	C_{44}	3.190457×10^{-6}
C_{31}	-7.982112×10^{-4}	C_{41}	6.394125×10^{-4}	S_{44}	5.514632×10^{-6}

1996 HW1 (all S_{nm} terms are zero)					
C_{20}	-1.21847×10^{-1}	C_{31}	-1.3964×10^{-2}	C_{41}	0
C_{21}	0	C_{32}	0	C_{42}	-4.258×10^{-3}
C_{22}	-5.8547×10^{-2}	C_{33}	2.547×10^{-3}	C_{43}	0
C_{30}	0	C_{40}	3.8779×10^{-2}	C_{44}	5.16×10^{-4}

J. Yuan et al., *Low Energy Flight: Orbital Dynamics and Mission Trajectory Design*, https://doi.org/10.1007/978-981-13-6130-2

Betulia					
C_{20}	-1.476131×10^{-1}	S_{31}	-2.491845×10^{-3}	S_{41}	-5.428366×10^{-4}
C_{21}	0	C_{32}	-5.879324×10^{-3}	C_{42}	-1.599034×10^{-3}
S_{21}	0	S_{32}	2.931994×10^{-5}	S_{42}	5.556629×10^{-5}
C_{22}	1.711891×10^{-2}	C_{33}	3.182376×10^{-4}	C_{43}	1.775273×10^{-4}
S_{22}	0	S_{33}	-3.910856×10^{-3}	S_{43}	2.49498×10^{-4}
C_{30}	9.543225×10^{-3}	C_{40}	4.2618×10^{-2}	C_{44}	-3.298214×10^{-5}
C_{31}	-2.738977×10^{-3}	C_{41}	-6.251823×10^{-4}	S_{44}	3.024807×10^{-5}

References

1. P. Tricarico, M.V. Sykes, The dynamical environment of Dawn at Vesta. Planet. Space Sci. **58**(12), 1516–1525 (2010)
2. J. Feng, R. Noomen, J. Yuan, Orbital motion in the vicinity of the non-collinear equilibrium points of a contact binary asteroid. Planet. Space Sci. **117**, 1–14 (2015)
3. S. Tzirti, H. Varvoglis, Motion of an Artificial Satellite around an Asymmetric, Rotating Celestial Body: Applications to the Solar System. PhD dissertation, Aristotle University of Thessaloniki, 2014
4. A. Konopliv, S. Asmar, R. Park, B. Bills, F. Centinello, A. Chamberlin, A. Ermakov, R.Gaskell, N. Rambaux, C. Raymond, The Vesta gravity field, spin pole and rotation period, landmark positions, and ephemeris from the Dawn tracking and optical data. Icarus **240**, 103–117 (2014)

Appendix C
The Location of EPs and Resonance Width

The location of EPs and resonance width at different combinations of *e* and *i* for Vesta, 1996 HW1 and Betulia

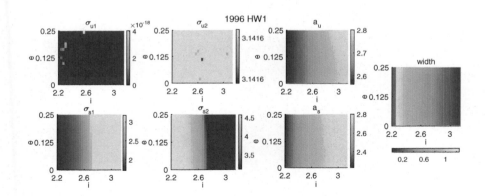

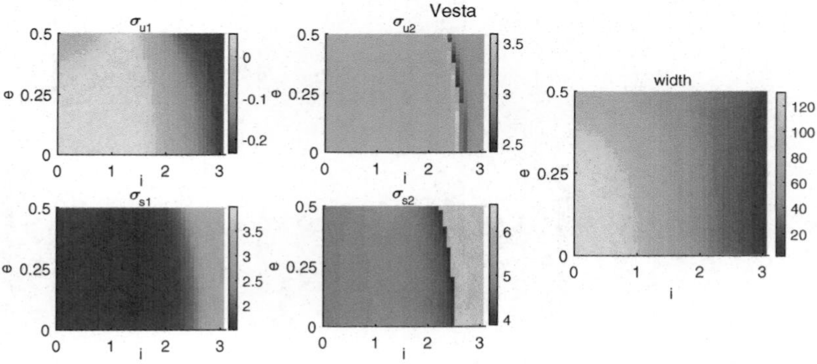

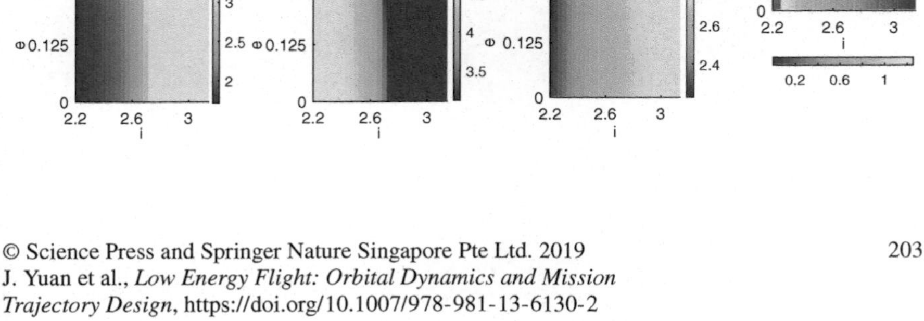

J. Yuan et al., *Low Energy Flight: Orbital Dynamics and Mission Trajectory Design*, https://doi.org/10.1007/978-981-13-6130-2

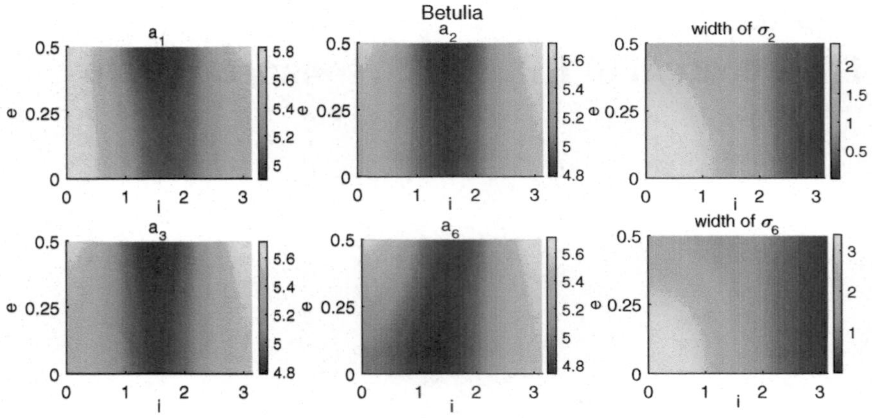
Betulia
a1
a2
width of σ2
a3
a6
width of σ6
e
i

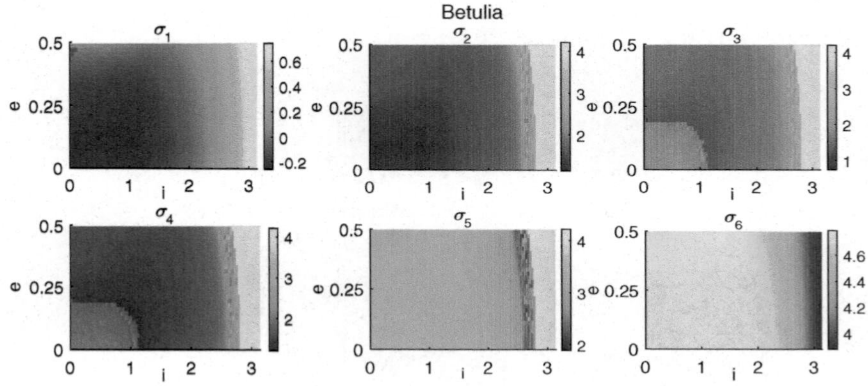
Betulia
σ1
σ2
σ3
σ4
σ5
σ6
e
i

Appendix D
The Second Derivatives of the Potential at the EPs Located at $(x_0, y_0, 0)$

$$
\begin{aligned}
\Omega_{11} &= 1 + \delta \left\{ \frac{3\mu R_1^2}{R_a^{5/2}} - \frac{\mu}{R_a^{3/2}} + \frac{3(1-\mu)R_2^2}{R_b^{5/2}} - \frac{1-\mu}{R_b^{3/2}} \right. \\
&\quad \left. + (1-\mu) \left[\begin{array}{l} C_{20}\left(-\frac{15R_2^2}{2R_b^{7/2}} + \frac{3}{2R_b^{5/2}} \right) + \frac{6C_{22}}{R_b^{5/2}} - \frac{60C_{22}R_2^2}{R_b^{7/2}} \\ + \frac{105C_{22}R_2^2R_3}{R_b^{9/2}} - \frac{15C_{22}R_3}{R_b^{7/2}} + \frac{C_{40}}{8}\left(\frac{105R_2^2}{R_b^{9/2}} - \frac{15}{R_b^{7/2}} \right) \end{array} \right] \right\} \\
\Omega_{12} &= \delta \left[\frac{3\mu\sigma_1 R_1 y_0}{R_a^{5/2}} + \frac{3(1-\mu)\sigma_2 R_2 y_0}{R_b^{5/2}} \right. \\
&\quad \left. + (1-\mu)\left(-\frac{15C_{20}\sigma_2 R_2 y_0}{2R_b^{7/2}} + \frac{105C_{22}R_3\sigma_2 R_2 y_0}{R_b^{9/2}} + \frac{105C_{40}\sigma_2 R_2 y_0}{8R_b^{9/2}} \right) \right] \\
\Omega_{22} &= 1 + \delta \left\{ \frac{3\mu y_0^2}{R_a^{5/2}} - \frac{\mu}{R_a^{3/2}} + \frac{3(1-\mu)y_0^2}{R_b^{5/2}} - \frac{1-\mu}{R_b^{3/2}} \right. \\
&\quad \left. + (1-\mu) \left[\begin{array}{l} C_{20}\left(-\frac{15y_0^2}{2R_b^{7/2}} + \frac{3}{2R_b^{5/2}} \right) - \frac{6C_{22}}{R_b^{5/2}} + \frac{60C_{22}y_0^2}{R_b^{7/2}} \\ + \frac{105C_{22}R_3y_0^2}{R_b^{9/2}} - \frac{15C_{22}R_3}{R_b^{7/2}} + \frac{C_{40}}{8}\left(105\frac{y_0^2}{R_b^{9/2}} - \frac{15}{R_b^{7/2}} \right) \end{array} \right] \right\} \\
\Omega_{33} &= \delta \left[-\frac{\mu}{R_a^{3/2}} - \frac{1-\mu}{R_b^{3/2}} + (1-\mu)\left(\frac{9C_{20}}{2R_b^{5/2}} - \frac{15C_{22}R_3}{R_b^{7/2}} - \frac{75C_{40}}{8R_b^{7/2}} \right) \right]
\end{aligned}
\tag{D.1}
$$

in which $R_3 = (x_0+\mu)^2 - y_0^2$, $R_a = (x_0+\mu-1)^2$, $R_b = (x_0+\mu)^2$, and $\sigma_1 = \text{sign}(x_0) + \mu - 1, \sigma_2 = \text{sign}(x_0+\mu)$.

J. Yuan et al., *Low Energy Flight: Orbital Dynamics and Mission Trajectory Design*, https://doi.org/10.1007/978-981-13-6130-2

Appendix E
The First and Second Derivatives of the ξ Component

The first and second derivatives of the ξ component.

$$
\begin{cases}
\dot{\xi} = \omega \dfrac{\partial \xi}{\partial \theta_1} + v \dfrac{\partial \xi}{\partial \theta_2} + u \dfrac{\partial \xi}{\partial \theta_3} & (E.1a) \\
\ddot{\xi} = \omega^2 \dfrac{\partial^2 \xi}{\partial \theta_1^2} + v^2 \dfrac{\partial^2 \xi}{\partial \theta_2^2} + u^2 \dfrac{\partial^2 \xi}{\partial \theta_3^2} + 2\omega v \dfrac{\partial^2 \xi}{\partial \theta_1 \partial \theta_2} + 2\omega u \dfrac{\partial^2 \xi}{\partial \theta_1 \partial \theta_3} + 2vu \dfrac{\partial^2 \xi}{\partial \theta_2 \partial \theta_3} & (E.1b)
\end{cases}
$$

At the left-hand side of the equation, the derivative terms at order come from two parts. One is from the frequency part at order N_f, and the other is from the partial derivatives part at order N_d. From Eq. (E.1a), the combinations of N_f and N_d are shown in Table E.1. Similarly, the combinations of N_f and N_d from Eq. (E.1b) are shown in Tables E.2 and E.3.

In the following tables, δ_{ij} is the Kronecker function defined as $\delta_{ij} = 0$ if $i \neq j$ and $\delta_{ij} = 1$ if $i = j$ (Table E.4).

Table E.1 The terms at order N from the first derivatives

N_f	N_d	$\omega \cdot \frac{\partial \xi}{\partial \theta_1}$	$v \cdot \frac{\partial \xi}{\partial \theta_2}$	$u \cdot \frac{\partial \xi}{\partial \theta_3}$
0	N	$-\omega_0 l \xi_{ijk}$	$-v_0 m \xi_{ijk}$	$-u_0 n \xi_{ijk}$
$N-1$	1	$-2b_1 \omega_{i-1jk} \delta_{l1} \delta_{m0} \delta_{n0}$	0	0

Table E.2 The terms at order N from the second derivatives

N_f	N_d	$\omega^2 \cdot \frac{\partial^2 \xi}{\partial \theta_1^2}$	$v^2 \cdot \frac{\partial^2 \xi}{\partial \theta_2^2}$	$u^2 \cdot \frac{\partial^2 \xi}{\partial \theta_3^2}$
0	N	$-\omega_0^2 l^2 \xi_{ijk}$	$-v_0^2 m^2 \xi_{ijk}$	$-u_0^2 n^2 \xi_{ijk}$
$N-1$	1	$-2\omega_0 \omega_{i-1jk} \delta_{l1} \delta_{m0} \delta_{n0}$	$-2v_0 v_{ij-1k} \delta_{l0} \delta_{m1} \delta_{n0}$	0

J. Yuan et al., *Low Energy Flight: Orbital Dynamics and Mission Trajectory Design*, https://doi.org/10.1007/978-981-13-6130-2

Table E.3 The terms at order N from the second derivatives

N_f	N_d	$\omega v \cdot \frac{\partial^2 \xi)}{\partial\theta_1 \partial\theta_2}$	$\omega u \cdot \frac{\partial^2 \xi}{\partial\theta_1 \partial\theta_3}$	$vu \cdot \frac{\partial^2 \xi}{\partial\theta_2 \partial\theta_3}$
0	N	$-\omega_0 v_0 lm \xi_{ijk}$	$-\omega_0 u_0 ln \xi_{ijk}$	$-v_0 u_0 mn \xi_{ijk}$
$N-1$	1	0	0	0

Table E.4 The coefficients of the third-order solution based on the system 1996 HW1 with $\delta = 30$

i, j, k	l, m, n	ξ_{ijk}^{lmn}	$\bar{\xi}_{ijk}^{lmn}$	η_{ijk}^{lmn}	$\bar{\eta}_{ijk}^{lmn}$
First order					
0 0 1	0 0 1	0	0	0	0
0 1 0	0 1 0	1	0	−0.035053982599398	−0.462561390510061
1 0 0	1 0 0	1	0	−0.039537507604550	−0.336524799123989
Second order					
2 0 0	0 0 0	−0.023806114920661	0	−0.053741389350147	0
	2 0 0	−0.016192214114086	0.035431401331960	−0.04459758386557	0.003759838927571
0 2 0	0 0 0	−0.021006982631613	0	−0.038383674187688	0
	0 2 0	−0.003915952890528	0.016657054924871	−0.043517860201724	−0.002029452661860
0 0 2	0 0 0	−0.000877747450392	0	−0.081689054179531	0
	0 0 2	−0.003434126850330	−0.078070935564490	−0.079513547812876	0.003416490374532
1 1 0	1 1 0	−0.011709442031593	0.039582384163479	−0.093135622475414	−0.002076374704935
	1 −1 0	−0.071866545748974	0.061834804112778	−0.102857914964174	−0.015387784859370
Third order					
0 1 2	0 1 0	0	−0.254455536743036	−0.118039682453026	0
	0 1 2	0.007553124943105	−0.000174979775855	−0.000184016696880	−0.008296739941255
	0 1 −2	0.030236354551945	−0.000380069835466	−0.000684739985558	0.025453086732806
1 0 2	1 0 0	0	0.258292868721213	0.085816387449906	0
	1 0 2	0.009817337527766	−0.000282319704210	−0.000302255508270	−0.010613532662720
	1 0 −2	0.018669336525655	0.000382335070366	−0.000672394037954	0.018179956333375
1 2 0	1 0 0	0	−0.023168973443996	0.005885433654049	0
	1 2 0	−0.009636398933641	−0.001840065809099	0.001846933760841	−0.013413088064179
	1 −2 0	0.008171128122171	0.002687712210761	0.006910586800193	0.004789674411975
2 1 0	0 1 0	0	−0.435139147907026	−0.185246194642733	0
	2 1 0	−0.010238386420191	−0.002425723361892	0.002825269628812	−0.015119405158079
	2 −1 0	0.047799298598753	−0.005380853876668	0.006962113291533	−0.020744877091795
0 3 0	0 0 0	0	−0.250160272880210	−0.110317571405433	0
	0 3 0	−0.003096643426563	−0.000554947829432	0.000540105182793	−0.004300085512299
3 0 0	0 0 0	0	0.049726518676773	0.023948768284819	0
	3 0 0	−0.003617421378978	−0.001217724734530	0.001544580406246	−0.006174181192942

i, j, k, l, m, n	ς_{ijk}^{lmn}	$\bar{\varsigma}_{ijk}^{lmn}$
	First order	
0 0 1 0 0 1	0	1
	Second order	
0 1 1 0 1 1	−0.001745243543444	0.092712440784877
0 1 1 0 1 −1	0.004267433115526	−0.22669852669720
1 0 1 1 0 1	−0.001466326438136	0.116825844612141
1 0 1 1 0 −1	−0.001745243543444	0.092712440784877
	Third order	
0 2 1 0 0 1	0	0
0 2 1 0 2 1	−0.010751558776951	−0.00048027511378
0 2 1 0 2 −1	−0.007691552844634	0.000570762759971
2 0 1 0 0 1	0	0
2 0 1 2 0 1	−0.014594281836922	−0.001378172317923
2 0 1 2 0 −1	−0.002785187193666	0.003328262370502
0 0 3 0 0 1	0	0
0 0 3 0 0 3	−0.000124241324214	0
1 1 1 1 1 1	−0.023981112829593	−0.001146204733178
1 1 1 1 −1 1	−0.035826340467038	−0.012350389128375
1 1 1 1 1 −1	−0.011862800003690	0.001773914178153
1 1 1 1 −1 −1	−0.001845253539161	0.009327034565701

i, j, k	ω_{ijk}	v_{ijk}	u_{ijk}
2 0 0	−0.020030813330247	0.024142612320952	0.001863199630957
0 2 0	−0.033184598445032	0.007740025281966	0.001878519971433
0 0 2	−0.026919160761714	0.019745386404717	−0.000506169348563